Augmented Reality and Its Application

Edited by Dragan Cvetković

Published in London, United Kingdom

IntechOpen

Supporting open minds since 2005

Augmented Reality and Its Application
http://dx.doi.org/10.5772/intechopen.95165
Edited by Dragan Cvetković

Contributors
Kedar Khare, Sarita Ahlawat, Purnima Sharma, Ankita Pandey, Durga Bisht, Apoorv Pant, Ritika Malik, Sandeep R Mathur, Aanisa Jan, Kiran Agarwal, Smita Singh, Meeta Singh, George Nehmetallah, Logan Williams, Thanh Nguyen, Shuva Chowdhury, Olika Moila, Andile Mji, Sibongile Simelane-Mnisi, Alex Pacheco, Kevin Sánchez, Rosario Pariona, Nestor Cuba, Bertha Larico, Silvia García, Paulina Trejo, Alberto García, Balakrishnan Chandrasekar, Matevž Dolenc, Angela Cristina De Hugo Silva, Metod Gaber, Ezgi Pelin Yildiz, Geoffreyjen Edwards, Dominique Michaud, Jonathan Proulx Guimond, Jonathan Caron-Roberge, Seppo Valli, Mika Hakkarainen, Pekka Siltanen

First published in London, United Kingdom, 2022 by IntechOpen
IntechOpen is the global imprint of INTECHOPEN LIMITED, registered in England and Wales, registration number: 11086078, 5 Princes Gate Court, London, SW7 2QJ, United Kingdom
Printed in Croatia

British Library Cataloguing-in-Publication Data
A catalogue record for this book is available from the British Library

Additional hard and PDF copies can be obtained from orders@intechopen.com

Augmented Reality and Its Application
Edited by Dragan Cvetković
p. cm.
Print ISBN 978-1-83969-704-3
Online ISBN 978-1-83969-705-0
eBook (PDF) ISBN 978-1-83969-706-7

For EU product safety concerns:
IN TECH d.o.o., Prolaz Marije Krucifikse Kozulić 3, 51000 Rijeka, Croatia, info@intechopen.com or visit our website at intechopen.com.

Meet the editor

Dragan Cvetković obtained a Ph.D. in Aeronautics from the Faculty of Mechanical Engineering, University of Belgrade, in 1997. To date, he has published sixty-five books, scripts, and practicums about computers and computer programs, aviation weapons, and flight mechanics. He has published many scientific papers as well. Dr. Cvetković became a full professor of Informatics and Computing at Singidunum University, Belgrade, in 2014. Since 2019, he has been the vice-rector for teaching at the same university.

Contents

Preface

Every time some new media appears, civilization accelerates in different directions. In the past, time seemed to move more slowly and was measured in weeks, days, hours, and minutes. Nowadays, it is measured in seconds, and surely there will come a time when it will be measured in milliseconds. We are currently living in the era of high-speed internet, laptops, smartphones, and watches, and we are rapidly approaching the moment when we will live in the time of a new medium, that of augmented and virtual reality.

Augmented Reality (AR) is a discipline that includes the interactive experience of a real-world environment, where real-world objects and elements are enhanced using computer perceptual information. AR adds elements of the virtual environment to the real world so that they look like part of it. This expands the user's view of the world with additional information that is directly embedded in the real world. In some applications it is not necessary to completely replace reality with a virtual world; sometimes it is only necessary to supplement or improve reality with some virtual parts.

AR can be defined as a system that incorporates three basic characteristics: a combination of the real and virtual world, real-time interaction, and precise verification and positioning of 3D virtual and real objects. AR changes the constant perception of the real-world environment, while virtual reality completely replaces it with a simulated one. AR is associated with two terms: Mixed Reality (MR) and Computer Mediated Reality (CMR).

AR is a relatively new area. Although the basic idea appeared in the beginning of the twentieth century, its rapid development didn't begin until the end of the same century; it has yet to achieve its full expansion. AR provides direct access to information and displays it in the user's field of vision, intertwining with the real world. This allows better, faster, and easier access to information. AR can be applied in medicine, manufacturing and maintenance, architecture, robotics, the military industry, and entertainment. When it comes to medical applications, AR can be used to overlay medical images with the patient, resulting in a kind of virtual X-ray that occurs in real-time. The result is that the doctor can see the patient's organs because the body is transparent. In production and maintenance, AR can be used to display visual instructions directly onto equipment/machinery, and thus the operator, instead of looking at the documentation, has all the necessary information at the right time in the right place. AR can be used in interior design to visualize structures or installations. For example, virtual furniture can be deployed in an actual room, thus one can get an impression of spatial relations and how the room will really look with the furniture. Military pilots can use AR to receive additional information and guidance as well as to see targets or guided missiles. In these cases, the AR display is built into the user's helmet or on the cabin.

AR is used to enhance the natural or real environment and offers perceptually enriched experiences. With the help of advanced AR technologies (computer vision, AR cameras in applications for smartphones and tablets, object recognition), information about the real world of users in the environment becomes interactive and can be manipulated digitally. It should be noted that AR has a lot of potential in the collection and exchange of knowledge.

This book is divided into two parts: "Augmented Reality in Education and Medicine" and "Augmented Reality and Engineering."

The first section consists of five chapters. Although AR applications are used in many areas, the most important of these areas is education. AR technology allows the combination of real objects and virtual information to increase students' interaction with physical environments and facilitate their learning. The first chapter in this section discusses how AR technology enables students to learn complex topics in a fun and easy way through virtual reality devices. Students interact with objects in the virtual environment and can learn more about them. For example, by organizing digital tours of a museum or zoo in a completely different country, lessons can be taught in the company of a teacher as if they were there at that moment. The second chapter is dedicated to the experiences of lecturers who used VR resources at a university in South Africa. The third chapter discusses cytopathology using high-resolution digital holographic microscopy. The fourth chapter presents the latest advances in digital holography with one or more wavelengths as well as holographic microscopy. The fifth chapter reviews the available AR or extended AR for technical vocational education training and skill development and its relevance in increasing the impact of student learning in skill centers.

The second section contains six chapters. VR and AR provide researchers, government authorities, and rescue teams with tools for recreating emergencies entirely through computer-generated signals of sight, sound, and touch (in VR), and overlays of sensory signals for experiencing a rich juxtaposition of virtual and real worlds simultaneously (in AR). The gap between knowledge and action is filled with visual, aural, and kinesthetic immersive experiences that make it possible to attend to the population in danger in a deeply efficient way, never experimented before. The first chapter in this section is dedicated to this issue. The benefits of AR technologies have been well proven in collaborative industrial applications, for example, in remote maintenance and consultancy. Benefits may also be great in telepresence applications, where virtual and mixed reality (nowadays often referred to as extended reality or XR) technologies are used for sharing information or objects over a network. The second chapter is dedicated to the advances in spatially true 3D telepresence. The third chapter provides an overview of building data modeling and the current state of the art in the use of augmented reality in various user scenarios of building data modeling and explores various challenges that need to be addressed for the adoption of AR technology in architecture, engineering, and construction in general. Virtual tools with their pre-set operability limit the designer's ways of interaction with artefacts. The fourth chapter proposes a framework for designers to interact with non-design experts through enhanced communicative media. The design framework indicates steps of design thinking to develop the interface by understanding both the virtual artefacts' perceptual affordance to the users and the design task. The chapter discusses projects tested in three different scenarios: urban design, architecture, and product design. Inspired by ideas portrayed in science fiction, the authors seek to develop a set of AR fashions that showcase scenes from a science fiction novel recently published by the principal author. The development team included artists and designers, a programmer, and the writer. Significant technical challenges needed to be overcome for success, including fabric construction and manipulation, image enhancement, robust image recognition and tracking capabilities, and management of lighting and suitable backgrounds. Viewing geometries were also a non-trivial problem. The final solution permitted acceptable but not perfect real-time tracking of the fashion models and the visualization of both static and dynamic 3D elements overlaid onto the physical garments. The fifth chapter is dedicated to this topic. AR is a technology that provides more interactive advertising, where you can manipulate and be part of it with greater clarity and empathy. In the digital era, companies in the retail sector face a new consumer profile that is more digital, more aware, and more informed. Therefore companies are in constant competition to impress their customers. The sixth chapter in this section describes the importance of AR in promoting products and services as an innovative alternative that captures attention and influences customer purchase decisions.

I would like to express my sincere gratitude to all the authors and co-authors for their contributions. TI would especially like to thank Publishing Process Manager Ms. Karmen Đaleta at IntechOpen for her support during the publishing process.

Dragan Cvetković
Singidunum University,
Beograd, Serbia

Section 1

Augmented Reality in Education and Medicine

Chapter 1

Augmented Reality Research and Applications in Education

Ezgi Pelin Yildiz

Abstract

Augmented reality is defined as the technology in which virtual objects are blended with the real world and also interact with each other. Although augmented reality applications are used in many areas, the most important of these areas is the field of education. AR technology allows the combination of real objects and virtual information in order to increase students' interaction with physical environments and facilitate their learning. Developing technology enables students to learn complex topics in a fun and easy way through virtual reality devices. Students interact with objects in the virtual environment and can learn more about it. For example; by organizing digital tours to a museum or zoo in a completely different country, lessons can be taught in the company of a teacher as if they were there at that moment. In the light of all these, this study is a compilation study. In this context, augmented reality technologies were introduced and attention was drawn to their use in different fields of education with their examples. As a suggestion at the end of the study, it was emphasized that the prepared sections should be carefully read by the educators and put into practice in their lessons. In addition it was also pointed out that it should be preferred in order to communicate effectively with students by interacting in real time, especially during the pandemic process.

Keywords: augmented reality research and applications, field of education, pandemic process, digital transformation, virtual environment

1. Introduction

Today, rapid changes and advances in science and technology affect and change the lifestyle of individuals. Apart from individuals, it is not possible for the education process and educational environments not to be affected by this change [1]. When the technologies used in educational environments from the past to the present are examined, it is seen that there is a transformation from blackboard and chalk to the computer and internet world, even to smart technologies with artificial intelligence. Especially in recent years, computer and internet technologies have had such a wide area of use in our lives that it was unthinkable for education services to be left out of the field [2].

The definition of today's learners as Z generation and/or digital generation and their characteristics require educators to follow technological developments and use the most appropriate technological tools in learning environments. One of these new technologies is augmented reality applications in education. When the literature is examined, there are many definitions of the concept of augmented reality made by researchers. Some of these definitions:

Augmented reality according to Milgram and Kishino [3]; "it is a reality environment where digital media products are used instead of real world objects" appears to be the most general definition. According to Azuma [4], augmented reality is a derivative of virtual reality. According to this definition, augmented reality is virtual environments in which existing reality is supported, not created from scratch. In this context virtual and real objects in augmented reality environments offered to users in harmony. Augmented reality creates the interactive environment between the virtual and real world. Augmented reality is used to achieve this [5, 6]. When the definitions in the literature are examined, as a common definition; augmented reality can be defined as real worlds enriched using virtual objects.

When the important areas where Augmented Reality (AR) Technology is used are examined;

- Education
- Health
- Marketing
- Game and Video
- Tourism
- Build
- Cinema
- Food
- Art and Museums
- Automotive
- Device Maintenance/Support

With the rapid development of Augmented Reality applications day by day, usage areas in many sectors are starting to increase. Major brands have started to give importance to providing a more realistic and embodied experience to their customers by using Augmented Reality (AR). This technology, which appears in many fields such as cosmetics, automobiles, construction, food, combines the virtual world with real life. Identifying target audiences, tracking and using technology in brand awareness and sustainable marketing is now vital for companies. The most importantly, companies from the public or private sector invest on enhanced technology in order to better promote or market their services/products and need talented people/firms in this field. In this context, augmented reality applications offer these services to businesses with technology support.

Although augmented reality applications are used in many areas, the most important of these areas is the field of education. New opportunities offered by AR technology for education have started to attract the attention of educators over time [7]. When these new opportunities and advantages are evaluated [8–11]:

- to provide students with more flexible and interesting learning environments,

- to experience an excitement they have never experienced before,
- to increase their willingness and motivation to learn,
- to help students make active observations during their learning processes and to form hypotheses as a result of these observations,
- to increasing students' learning performance and helping them establish social interactions within the group,
- to bridging formal and informal learning and encouraging students to learn collaboratively,
- AR technology; it gives a feeling of independence from the place, freedom and personal,
- to creating new opportunities in education by promoting learning.
- it is possible to rank as.

When the augmented reality technologies, which are frequently used in the field of education, are examined, **wearable technologies** draw attention. Wearables are loaded with smart sensors that track body movements. Usually these products use bluetooth, Wi-Fi and mobile internet connection to sync with smartphone wirelessly. Users are connected to wearable devices with the help of sensors. Wearable technology products that are always with the user; it provides important services in many areas, especially in entertainment, health, work, information, education, socialization and security.

Wearable technologies in the field of education are used in learning-teaching environments. Modern visualization techniques help students explore existing educational resources and new knowledge (**Figure 1**) [12].

Wearable technologies frequently used in education:

- Internet of things
- Smart watches
- Google – Glass Project

Figure 1.
Wearable technologies the past and present and future.

- HoloLens – Microsoft:
- Oculus Rift – Facebook
- Bracelets, Rings and Necklaces
- Smart Clothing and Tattoos

These tools, which can also be named as wearable computers in the literature, reveal a commensalistic relationship between human and computer however, the daily life of the individual has a structure that enriches their experience [13]. From smart watches to wristbands, sensor accessories such as rings and necklaces, virtual reality glasses, Google Glass project and derivative smart glasses, as well as smart optical lenses and headphones, many things can be shown among wearable technologies [14].

When the programs that enable the use of AR technologies in education are considered:

- Augment – 3B
- Google Translate
- SketchAR
- Wikitude
- LifePrint Photos
- Smartify
- Spyglass
- Blippar
- Aurasma

In the light of all this information, the purpose of this chapter; the use of augmented reality environments and applications in the field of education, the programs and technologies used in this context, and the researches are discussed in detail.

The new normal situation, especially with the pandemic process, also creates an opportunity for more educators to try new generation technologies (VR and AR technologies) beyond video and teleconferencing applications. It is predicted that such research studies will be important so that educators realize the benefits of these technologies and use them actively in learning environments.

2. Conceptual framework

Augmented reality (AR) has been slowly but surely following its predecessor virtual reality in changing the education sector—digitizing classroom learning, and making training more diverse and interactive. In this section, current studies in the literature in recent years on the integration of augmented reality applications into education are given. When these studies are examined;

Çetin [15], investigated the effect of augmented reality-based stories on reading skills in his research. In the research, augmented reality based story text samples were presented to primary school 3rd grade students (**Figure 2**).

A scoring key was developed for the answers given to the questions prepared by the researcher to measure the skills of expressing what they read in writing. As a result of the research, it was observed that the augmented reality-based stories did not have a significant effect on the reading motivation and reading comprehension skill levels of the students, but they created a positive significant difference on their ability to tell what they read in written and verbal form. In addition, as a result of the research, it was observed that the reactions of the students towards the texts increased.

As a similar study Baysan and Uluyol [16], the effect of the use of augmented reality books (AR-books) on the academic success of the students and the students' opinions about the environment were investigated in his study. The AR-based teaching material developed by the HITLibHZ-BuildAR program was used in the laboratory environment for the experimental group of 22 people and the course was taught by the researcher. As a result; according to the qualitative data obtained from the students, AR is a promising technology. Educational AR applications should be used in areas that require 3D spatial visualization such as Geometry and Geography rather than technology education. Participants support the use of AR in Computer Hardware training, with better developed platforms and more professional designs (**Figure 3**).

Almusawi et al. [17], in their study, they discussed innovation in physical education: teachers' perspectives on readiness for wearable technology integration. The study is a case study and includes semi-structured interviews with 38 public school physical education teachers. The following scheme was used in the study (**Figure 4**).

Figure 2.
Augmented reality based story text samples.

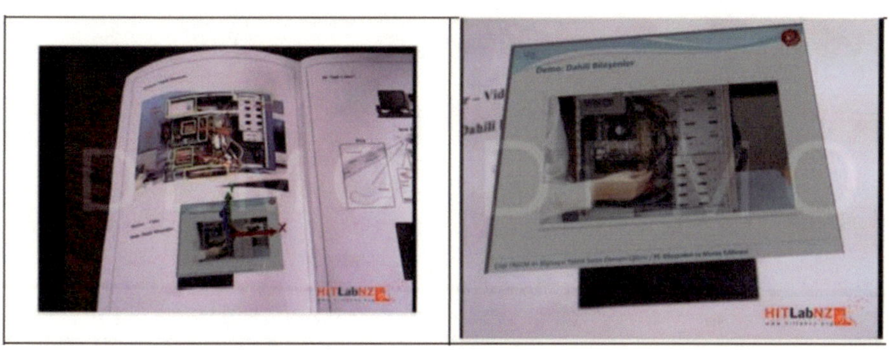

Figure 3.
Augmented reality application book sample.

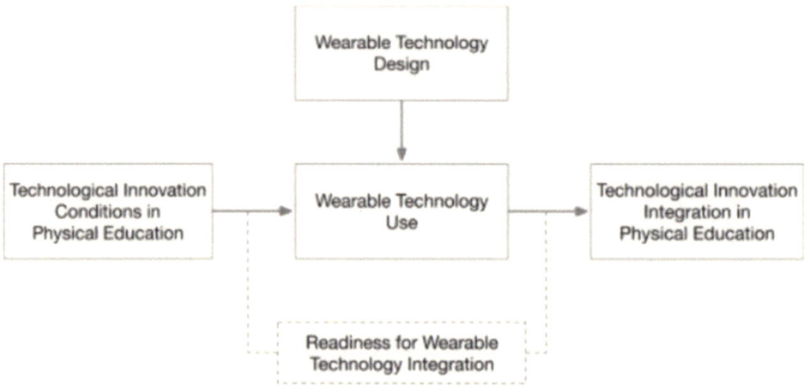

Figure 4.
Augmented reality application book sample.

The findings show that physical education teachers have concerns about the design aspects of wearable technologies in terms of material design and device suitability for physical education. To eliminate these concerns, it is proposed to provide innovative learning environments that impact technology through collaborative, competitive, engaging and evidence-based learning experiences through wearable technologies that provide comfort, enhanced wearability and injury prevention in physical education.

It is understood from the existence of studies in the literature that augmented reality technologies have been used frequently in medical education recently. When the relevant studies in the literature are examined (**Figure 5**).

Kucuk et al. [18], a new perspective in medical education multimedia applications: augmented reality has been studied in their research. As a result, it is difficult to understand the subjects including the structure of the brain and vessels such as neuroanatomy in medical courses, in this direction, it was emphasized that AR applications could be developed to facilitate the learning processes of students in such subjects. Considering the characteristics of today's students in the digital citizen group, it has been suggested in the study that students should be supported

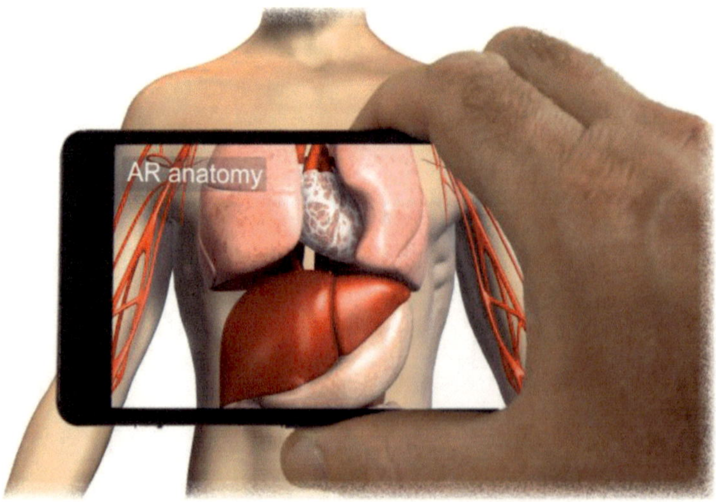

Figure 5.
Use of augmented reality technologies in medical education.

with various technological solutions in this process, at this point, the dissemination of medical augmented reality applications that are based on the learning approach anytime and anywhere and support individual learning.

3. Augmented reality applications used in education

Augmented reality, a concept that has been frequently encountered recently, promises a future where we can get away from the world we live in, create a new worlds and enter 'inside' our imagination. By adding this technology with which we can 'beautify' the world we live in, make brand new additions to our world and bring our imagination to the place we live in, we started to manipulate our real world at the same time, while constructing mixed reality virtual worlds that we use together. It has become compulsory to benefit from these privileges and advantages that augmented reality offers to our lives, especially in terms of education, on behalf of the Z generation youth.

It is now possible to use these technologies in learning and teaching environments by making use of the ready-made programs of augmented reality. When the literature is examined, the frequently used programs and application areas are below:

3.1 Augment: 3B

Augment is an ARCore-based mobile app to visualize 3D models in Augmented Reality, integrated in real time in their actual size and environment. Balak and Kısa [19] investigated the effects of this application on technical drawing education in their studies. The data obtained as a result of the use of Augmented Reality technology in the technical drawing course of the 2015–2016 period were examined. As a result; the result of the survey made with the pre- and post-tests applied; it has been determined that the students understand and adopt the Augmented Reality technology, which is a modern education tool, and this technology increases their interest in the lesson (**Figure 6**).

3.2 Google translate

According to Google, the Translate app currently supports text translations between 103 languages, offline translations for 52 languages and Word Lens-based augmented reality translations for 30 languages. Aiming to make life easier for users with its mobile translation application, Google offers Instant camera translation; It started to support a total of 88 languages with the addition of 60 new languages such as Arabic, Hindi, Malaysian, Thai and Vietnamese etc. (**Figure 7**).

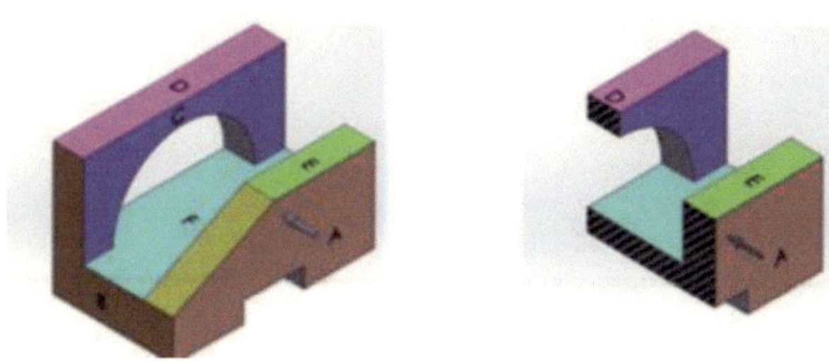

Figure 6.
Technical drawing with 3D modeling with AR technologies.

Figure 7.
Augmented reality-based Google translate app.

3.3 SketchAR

SketchAR, which is an application that combines augmented reality and drawing, is among the applications frequently preferred by artists recently. SketchAR, which is basically a drawing application made available to artists, confirms that digital works created by artists are unique and original, making them accepted as NFT (data unit). SketchAR, an initiative founded in 2017 by Aleksandr Danilin, Alexander Danilin and Andrey Drobitko in Lithuania, offers its users a different drawing experience by combining augmented reality technology with drawing, together with artificial intelligence support (**Figure 8**).

3.4 Wikitude

Wikitude initially focused on providing location-based augmented reality experiences through the Wikitude World Browser App. In 2012, the company restructured its proposition by launching the Wikitude SDK, a development framework utilizing image recognition and tracking, and geolocation technologies. Wikitude initially entered the market with its geo location AR app. The Wikitude

Figure 8.
Drawing courses with SketchAR.

Figure 9.
Wikitude world browser app.

app was the first publicly available application that used a location-based approach to augmented reality (**Figure 9**).

It is supported by studies in the literature that this application is also used in geography education. Wikitude; it is a complete AR development platform used by major brands, travel catalogs, retailers and publishers to deliver a variety of engaging solutions.

3.5 LifePrint photos

Life Print is an Android and iPhone photo and video printer. The Life Print program uses augmented reality to magically bring photos to life (**Figure 10**).

3.6 Smartify

The application starts with permission from users to access camera and location. With camera access, the artwork is scanned, and according to the location, it provides the opportunity to get information about which museums are and how far, how many artworks of art they are, open and closed hours, and to see some of the artworks in the museum. The application has three basic directions; *scan*, *profile* and *explore* (**Figure 11**).

Figure 10.
Augmented reality app: LifePrint photos.

Figure 11.
Augmented reality app: Smartify.

3.7 Spyglass

Spyglass app is a program that allows users to turn their smartphones into a compass, gyroscope, star tracker and more (**Figure 12**).

3.8 Blippar

Blippar uses augmented reality, artificial intelligence and computer vision to provide you with information about what you find around you. It is quite successful with its advanced image recognition algorithms that find out what the objects are and bring the relevant information. Blippar will introduce the feature that will allow its users to create their own profiles very soon, but it will be possible to get detailed information about a person with the innovation called Augmented Reality Face Profiles (**Figure 13**).

Figure 12.
Locating with spyglass technologies.

Figure 13.
Unlock augmented reality of everyday objects and places with the Blippar app.

3.9 Aurasma

One of the web 2.0 tools using Augmented Reality technology is the Aurasma application. Interactive virtual reality materials can be created free of charge with the Aurasma web 2.0 tool. With these materials, students can be taught more efficiently, and very effective information can be provided outside the classroom [20]. *How to Use Aurasma Web 2.0 Tool in Education*:

- by creating animated and interactive boards
- prepare interactive lecture notes or handouts
- interactive presentation of albums or details about activities such as observation projects, experiments (**Figure 14**).

According to Onder [21], the Aurasma application draws attention with its ability to provide AR environments and opportunities to teachers and students, ease of

Figure 14.
Educational use of Aurasma app.

use, support for distance education, creating individualized learning environments and being used as an evaluation tool.

4. Method

This research is an example of a literature review. A literature review is a search and evaluation of the available literature in your given subject or chosen topic area [22]. At the end of the study, it was emphasized that the prepared sections should be carefully read by the educators and put into practice in their lessons. In addition it was also pointed out that it should be preferred in order to communicate effectively with students by interacting in real time, especially during the pandemic process.

5. Conclusion and suggestions

In this research, a detailed analysis of the augmented reality environments and applications that are frequently used in the design of learning and teaching environments in the education sector with the digitalization process is included. As the general results of the research; today, with the introduction of technologies into educational environments, different tools and materials have begun to be used in teaching methods. In this context, it is seen that the inclusion of mobile tools and mobile applications in learning environments has become widespread recently. With this rapid development in mobile technologies, new media environments, in which interactivity increases, offer an increasing number of services to the user. One of the environments where this interaction is provided and which can integrate objects in virtual environments with real objects is technologies that offer "Augmented Reality (AR)". These technologies allow virtual objects to be superimposed on real images. AR tools consist of camera, computer infrastructure, a marker and tangible objects.

One of the most important sectors in which augmented reality technologies are used is the education area. Augmented reality applications help students understand abstract concepts in the learning and teaching process; it provides environments where students can share information within the group. In addition, it has been supported by studies in the literature that these environments significantly increase students' learning. In addition, it was emphasized that augmented reality increases the interests, motivations and experiences of students in the field of education and plays a role in transferring the knowledge and skills gained in the virtual environment to real environments.

In all this context; increasing the use of learning environments of augmented reality environments and applications, where the effectiveness of its use in education has been determined to this degree, in different levels and course contents is the most important suggestions of this research.

DOI: http://dx.doi.org/10.5772/intechopen.99356

Author details

Ezgi Pelin Yildiz
Department of Computer Programming, Kazim Karabekir Vocational School of Technical Sciences, Kafkas University, Kars, Turkey

*Address all correspondence to: yildizezgipelin@gmail.com

References

[1] Akkoyunlu, B. (2002). Eğitimde İnternet Kullanımı. İstanbul: Ceren and BİTAV.

[2] Bulun, M., Gülnar, B., and Güran, M. S. (2004). Eğitimde Mobil Teknolojiler. The Turkish Online Journal of Educational Technology, 3(2), 165-169.

[3] Milgram, P., Takemura, H., Utsumi, A. and Kishino, F. (1994). Augmented reality: A class of displays on the reality-virtuality continuum. Telemanipulator and Telepresence Technologies, SPIE, 2351, 282-292. DOI:10.1117/12.197321

[4] Azuma, R. (1997). A survey of augmented reality. Presence: Teleoperators and Virtual Environments, 6, 355-385. DOI:10.1162/pres.1997.6.4.355

[5] Bronack, S., Sanders, R., Cheney, A., Riedl, R., Tashner, J., and Matzen, N. (2008). Presence pedagogy: Teaching and learning in a 3D immersive world. International Journal of Teaching and Learning in Higher Education, 20(1), 59-69

[6] Klopfer, E. and Sheldon, J. (2010). Augmenting your own reality: Student authoring of science-based augmented reality games. New directions for youth development, 128, 85-94.

[7] Erbas, C. and Demirer, V. (2015). Eğitimde Sanal ve Arttirilmiş Gerçeklik Uygulamaları. In book: Eğitim Teknolojileri Okumaları 2015 Chapter: 7

[8] Chiang, T.H.C., Yang, S.J.H. and Hwang, G.J. (2014). An augmented reality-based Mobile learning system to improve students' learning achievements and motivations in Natural science inquiry activities, Educational Technology and Society, 17(4), 352-365.

[9] Ibáñez, M. B., DiSerio, Á., Villarána, D. and Kloosa, C. (2014). Experimenting with electromagnetism Using augmented reality: Impact on flow student experience and educational effectiveness, Computers and Education, 71, 1-13.

[10] Lin, T. J., Duh, H. B. L., Li, N., Wang, H. Y. and Tsai, C. C. (2013). An investigation of Learners' collaborative knowledge construction performances and behavior patterns In an augmented reality simulation system, Computers and Education, 68, 314-321. 7

[11] Wu, H. K., Lee, S. W. Y., Chang, H. Y. and Liang, J. C. (2013). Current status, opportunities and challenges of augmented reality in education, Computers and Education, 62, 41-49.

[12] Information Technology and Communication Agency Report (2020). Retreived from Information and Communication Technology in Special Needs Education report | European Agency for Special Needs and Inclusive Education (european-agency.org)

[13] Sezgin, S. (2016), "Eğitimde Giyilebilir Teknolojiler: Fırsatlar ve Eğilimler", Education Faculty Journal of Mehmet Akif Ersoy, 40, 405-418.

[14] MacLean, D. (2013). MoodWings: A Wearable Biofeedback Device for Real-Time Stress Intervention. Proceedings of the 6th International Conference on Pervasive Technologies Related to Assistive Environments (p. 66). ACM.

[15] Çetin, S. (2020). Artırılmış gerçeklik uygulamalarının teknik resim dersinde ortaöğretim öğrencilerinin akademik başarıları, tutumları ve uzamsal görselleştirme becerilerine etkisi. University of Bursa Uludag Turkey.

[16] Baysan, E, Uluyol, Ç. (2016). Arttırılmış Gerçeklik Kitabının (AG-KİTAP) Öğrencilerin Akademik Başarılarına Etkisi ve Eğitim Ortamlarında Kullanımı Hakkında Öğrenci Görüşleri. Journal of Education and Humanities: Theory and Practice, 7(14), 55-78.

[17] Almusawi HA, Durugbo, C., Afaf, M. (2021). Innovation in physical education: Teachers' perspectives on readiness for wearable technology integration. Computers and Education 167, DOI:10.1016/j.compedu.2021.104185

[18] Kucuk, S., Kapakin, S. ve Goktas, Y. (2015). Tıp fakültesi öğrencilerinin mobil artırılmış gerçeklikle anatomi öğrenimine yönelik görüşleri. Journal of Higher Education and Science, 5(3), 316-323.

[19] Balak, MV, Kısa, M. (2016). Artırılmış Gerçeklik Teknolojisinin Teknik Resim Eğitimi Üzerindeki Etkilerinin Arastırılması, Harran University Journal of Engineering, 2

[20] Squire, K. D. and Jan, M. (2007). Mad City mystery: Developing scientific argumentation skills with a place-based augmented reality game on handheld computers. Journal of science education and technology, 16(1), 5-29.

[21] Onder, R. Eğitimde artırılmış gerçeklik uygulamaları: Aurasma ve Color Mix. (2017) Retreived from https://docplayer.biz.tr/105856887-Egitimde-artirilmis-gerceklik-uygulamalari-aurasma-ve-color-mix.html

[22] Arshed, N., and Danson, M. (2015). The literature review. In R. MacIntosh, and K. D. O'Gorman (Eds.), Research Methods for Business and Management: A Guide to to Writing your Dissertation (Pp. 31-49). Goodfellow Publishers.

Chapter 2

Lecturers' Experiences in Teaching Using VR Resources at a Selected University, South Africa

Olika Moila, Andile Mji and Sibongile Simelane-Mnisi

Abstract

Globally, academics have reconnoitered the various benefits of virtual reality (VR) in education. This study explored the lecturers' experiences with VR resources in teaching and learning environments with pre-service teachers at one of the selected universities in South Africa. The study engaged a qualitative method, comprising of semi-structured face-to-face interviews with 6 lectures. The data from the interviews were evaluated by hand and the findings from this study were precisely described as given by the interviewees. These findings specified that the lecturers acknowledged the effectiveness of the use of VR resources in teaching and learning since all activities become more concrete., However, the lecturers indicated that there were some challenges that hindered them from employing VR resources into their teaching and learning environments and these included a lack of adequate lecturer development for the use of VR tools for teaching; inadequate VR tools for teaching and learning in their departments; VR resources were not tailor-made for their current curricula; and inadequate funding for 4IR resources. Hence, this study recommended that this university should immediately provide all the support required to facilitate the lecturers' use of VR resources for teaching to avoid the use traditional teaching and learning methods.

Keywords: virtual reality resources, challenges, traditional teaching and learning methods, ripple effect

1. Introduction

Fourth industrial revolution tools are being increasingly integrated into teaching and learning in education [1]. Globally, people are exploring the potential of these emerging technologies, which among many other mobile devices include virtual realities (VR) [2–4]. Furthermore, research also maintains engagement with VR enhances the experience or a sense of 'being there' in the environment [5]. Virtual reality enhance knowledge of abstract concepts, thus make impossible tasks to undertake in the real world possible [6]. In other words, VR allow participants to experiment with objects virtual more easily 'making the unseen seen', machines [7].

Even though VR is not new, the recent advances in immersive technologies in terms of visualization and interactions have made VR increasingly attracted attention by researchers. The use of VR enhances immersion of a user in a virtual environment, thus providing a sense of "being" in the task environment instead.

The immersion of a being in a virtual reality as the perception of being physically present in a non-physical world through the creation of images, sound, or other inducements so that a participant feels he or she is actually "there" [8]. However, this is not the case at the selected departments of university involved in this study as the pre-test study on the use of 4IR tools to equip pre-service teachers indicated ineffective preparation of students for the 4IR world. It is in this regard that the researchers embarked on this study as an intervention in order to enhance the lecturers' experiences with VR resources to enable pre-service teachers bridge the physical distance in teaching and learning environments at one of the selected universities, South Africa.

2. Review of literature

The general concept of immersive VR was developed back in the late 80s in which participants interact with a world completely generated by [9]. One of the main characteristics of VR is that the environment is a full scale replica of the real world and it relates to human size. Hence, the participants get the feeling as if they are interacting with the real environment or subject, and that VR applications include either real or abstract worlds [10]. In addition, through engaging in virtual reality, we can enter and interact with a world that either does not exist or it is difficult to access due to costs or safety reasons [9].

Virtual reality has been described as one of the most effective and necessary teaching and learning resources of the 4th industrial revolution [11]. Moreover, research argues that students tend to retain more information and gain more practical skill through their engagement with VR tools [12]. Virtual reality allows us to scrutinize all matter through touching, moving and manipulating it [13]. In other words, the use of VR allows participants to be totally engrossed in a self-contained artificial or simulated environment while experiencing it as real [14]. Hence, these tools can offer rich and complex content-based learning, as well as enhance students' technical, creative, and problem-solving skills [14].

A substantial body of researchers acknowledge that there are many potential benefits of introducing VR in education and training that would not be possible in traditional methods, and that these include: education which is not possible in reality, will be possible in virtual reality; virtual game-based experience increases students' motivation; collaboration in virtual reality classroom fosters social integration of learners; learning is achieved by direct interaction, not by mouse clicks; the results from the learning process are truly assessed [15–17].

Moreover, the use of VR is said to have advantages in teaching and learning that are not experienced when traditional pedagogy is practiced for delivering curricula [18]. This researcher points out that these advantages include enhanced motivation on the part of students; improved communication between lecturers and students; more comprehension of seemingly abstract concepts; and more individual learning needs are accommodated [18].

Despite the substantial benefits that the use of VR tools are said to have in the classroom, educational institutions are not free of significant challenges. Due to financial constraints many educational institutions are reluctant to invest in fourth industrial tools [19]. Furthermore, one of the biggest challenges faced by the use of these tools, including VR resources is the lack of relevant content. Developing more content can be very costly, and not every educational institute has the means to hire a software development company to help them produce content [19].

Echoing similar sentiments, it has been noted that cost is the primary challenge of 4IR tools, in this case VR tools in most markets, and especially so in education, hence funding is one of the primary barriers for massive adoption of VR solutions within educational systems, due to the high cost of relevant tools as there are limited financial resources in many educational institutions [20].

Moreover, knowledge about how to integrate VR technology in courses and the lack of time for learning and planning how to do this has also been identified as a challenge for pedagogical VR use at university level [21]. It has also been highlighted that opportunities to enact VR technologies is limited due to shortcomings in practitioners' continuing professional development opportunities [22].

3. Research methodology

This study used a qualitative research methodology to explore the experiences of 6 lecturers on their use of VR resources to teach pre-service teachers. Qualitative researchers assume that everyone views the world in their own way [23]. Hence, a qualitative approach was preferred for the reason that it shows a great potential value to this study for the researcher to get comprehensive and rich data on the experiences of lecturers in the use of VR resources for teaching. Since a qualitative approach assumes that the world cannot be viewed as a single reality, it was suitable to be used in this study because the way the participants give different accounts of their experiences on the use of VR resources depended on their natural setting. In this study, convenience sampling was used to select the participants. In this regard, convenience sampling is a type of sampling that involves the selection of the most accessible subjects, it is the least costly to the researcher, on terms of time, effort and money [24]. Hence, the researchers engaged face-to-face interviews with the six selected lecturers from various departments at the faculty of humanities in order to elucidate responses on their experiences with VR resources in teaching activities. In this qualitative study, the researchers were obliged to ensure that the research findings were credible and trustworthy so that they could be interpreted, applied in the field and benefits researchers and other interested parties.

Validity in qualitative research refers to the degree of resemblance between the explanation of the phenomenon and the realities of the situation in the given contexts [25]. Hence, in this study the four trustworthiness criteria which include credibility, transferability, dependability and confirmability were applied [26]. These criteria and their peculiar activities were adhered to in this study. Prior to the study, the researchers ensured that all the participants were informed about the purpose of the study. Participation in the study was voluntary as participants were informed of their rights to withdraw from the study at any time. Consent forms were be to participants to confirm or decline their participation in this study. Confidentiality and anonymity was ensured by using avoiding participants' real names.

4. Results and discussion

The demographic information of the lectures who participated in this study was elicited through the first part of the interview schedule. This was done in order to better contextualize the results. **Table 1** below represents the profiles of the participants:

The study reconnoitered the lecturers' views under the following themes which emerged from their responses:

Gender	Female		Male	
	2		4	
Age	20–30 years	31–40 years	50 years and above	
	2	3	1	
Level taught	Level 1	Level 2	Level 3	
	1	3	2	
Lecturing Experience at university	1–5 years	6–10 years	11–15 years	16 years and above
	2	2	1	1

Table 1.
Profiles of the lecturers.

5. The adequacy of VR resources to equip students with required skills

The lecturers were requested to explicate their experiences on whether or not there were adequate VR resources to equip pre-service teachers with required skills? Some of the lecturers' views are presented below:

- The department does not have adequate VR resources to use during and outside lecture rooms;
- If lecturers do not even have the resources, you can imagine the challenge relating to students' access to VR resources;
- VR resources are inadequate;
- I have not seen any except the personal smart phones that we have, but absence of Wi-Fi makes it impossible to use VR resources in classes;
- The department is not yet ready for the use of VR resources for teachers;
- There are not adequate VR resources, our line managers must procure for all of us.

Although globally, 4IR tools are increasingly being integrated into teaching and learning in education [1], the findings in theme 1 above indicate that this is not the case at the institution involved in this study. The lecturers stress that there is a lack of adequate VR resources, hence they are not yet ready to use these tools in their teaching and learning environments.

6. Benefits of using VR resources for teaching

The researchers explained to the participants how VR resources work (google expedition sites and googles), as well as allowed them to experience their use. Lectures were then requested to indicate any experienced benefits of using VR resources for teaching. In this regard, the following are some of the responses from the interviews:

- VR resources have great effect as the add to the current conversations of 4IR;
- Learning becomes concrete through the use of VR resources; and the interaction would arouse interest in students because of the current generation being the digital-age generation
- VR resources help to bring real-life realities into the classroom
- These tools eliminate challenges as students are able to see or view real-life videos in which concepts are clarified easily
- I think VR resources have a positive effect as they enhance knowledge in teaching and learning
- Through the use of VR tools, the students will be able to relate their academic work to the outside real world

The lecturers' perceptions on the benefits of using VR resources for teaching are supported by a substantial body of researchers who acknowledge that there are many potential benefits of introducing VR in education and training that would not be possible in traditional methods, and that these include: education which is not possible in reality, will be possible in virtual reality; virtual game-based experience increases students' motivation; collaboration in virtual reality classroom fosters social integration of learners; learning is achieved by direct interaction; and the results from the learning process are truly assessed [15–17].

Moreover, knowledge about how to integrate VR technology in courses and the lack of time for learning and planning how to do this has also been identified as a challenge for pedagogical VR use at university level. It has also been highlighted that opportunities to enact VR technologies is limited due to shortcomings in practitioners' continuing professional development opportunities [22].

7. Challenges of using VR resources for teaching

Lecturers were requested to elucidate their perceptions on the challenges of using VR tools for teaching and learning. In response, the following are some of the views from the interviews with the concerned lecturers:

- Like any other learning resources, VR resources have their own vocabulary and failure for the individuals to understand their operational aspect, therefore challenges would persist;
- The use of VR might be a problem to implement them if these resources are not tailor-made for the current curricula;
- These resources may only be used effectively if they are compatible with the users' abilities, otherwise trying to use them in the existing syllabi would not be effective;
- A lot of adaptability needs to be done for these tools to make a positive impact in education;

- Issued relating to professional development, funding and curricula alignment to the use VR tools need urgent consideration;

- Affordability and the feasibility of procuring of VR tools for all subjects need to be well planned for, otherwise we will not be able to use these resources anytime soon.

As indicated in lecturers' perceptions on the challenges of using VR tools for teaching and learning, operating with the realities of tight budgets and competing demands for funds, many educational institutions are reluctant to invest in what seem like expensive gadgets with no immediate tangible benefits [24]. Furthermore, he adds that one of the biggest challenges faced by VR education is the lack of content. The fact is that developing more content can be very expensive, and not every educational institute has the means to hire a software development company to help them produce content [19].

8. The influence that the use of VR has in the enhancement of teaching and learning compared to the traditional methods

The lecturers were asked to elucidate their experiences on whether or not the use of VR has any influence in the enhancement of teaching and learning compared to the traditional methods. Some of the lecturers' views are presented below:

- Yes, the VR resources may enhance teaching and learning only if the subject content is well researched and well aligned to the curriculum;

- Since these resources enable students to be practically engaged, they should enhance teaching and learning in a better way compared to the traditional methods;

- Yes, since the VR resources bring the world to the teaching and learning environment, they enhance teaching and learning activities more than traditional methods;

- The use of VR resources responds to the needs of the students, who are digitally savvy;

- VR tools enhance teaching and learning since students will learn through seeing and touching, hence become totally in control of their learning;

- VR resources give more meaning to learning content since abstract concepts can explained in a simpler manner than in traditional methods.

The lecturers' responses to whether or not the use of VR has any influence in the enhancement of teaching and learning compared to the traditional methods are in line with views echoed in literature. Moreover, the use of VR is said to have advantages in teaching and learning that are not experienced when traditional pedagogy is practiced for delivering curricula [18]. These advantages include enhanced motivation on the part of students; improved communication between lecturers and students; more comprehension of seemingly abstract concepts; and more individual learning needs are accommodated [18].

9. The possible solutions to any perceived challenges that hinder VR tools from being used in some classes

The lecturers were asked briefly provide their own views of the solutions to any perceived challenges that hinder VR tools from being used in some classes. Some of the lecturers' views are presented below:

- All lecturers in the department urgently need more training on the use of these resources;
- I personally need to catch up with latest trends but it starts with training and procurement of VR resources which are still at a very low level now;
- There is need for intensive lecturer development in the use of VR resources;
- All lecturers and students must be provided with VR tools and Internet access;
- The department must provide adequate VR tools and equipment, as well as align curricula to the use of resources such as these;
- There should be adequate resources and training, as well as continuous support on the use of these modern tools for teaching and learning.

The lecturers' views on the solutions to any perceived challenges that hinder VR tools from being used in some classes are supported in research which indicates that knowledge about how to integrate VR technology in courses and the lack of time for learning and planning how to do this is a challenge for pedagogical VR use at university level [21]. It has also been highlighted that opportunities to enact VR technologies is limited due to shortcomings in practitioners' continuing professional development opportunities [22]. Hence, the effective use of VR tools in teaching and learning environments can only be realized once the said challenges are dealt with.

10. Conclusions and recommendation

The findings indicated there were various benefits that VR resources have in education, and these included the following: 1. they add to the current conversations of 4IR; 2. learning becomes concrete through their use and the interaction would arouse interest in students because of the current generation being the digital-age generation; 3. They help to bring real-life realities into the classroom; 4. they eliminate challenges as students are able to see or view real-life videos in which concepts are clarified easily; 5. they have a positive effect as they enhance knowledge in teaching and learning; 6. VR resources may enhance teaching and learning only if the subject content is well researched and well aligned to the curriculum; 7. VR resources have the potential to enhance teaching and learning in a better way compared to the traditional methods; 8. VR resources bring the world to the teaching and learning environment hence, they enhance teaching and learning; 9. the use of VR resources responds to the needs of the students, who are digitally savvy; 10. VR tools enrich teaching and learning since students learn through seeing and touching, hence become totally in control of their learning; 11. VR resources give more meaning to learning content since abstract concepts can explained in a simpler manner than in traditional methods; and 12. Students are able to relate their academic work to the outside real world.

However, the findings also indicated that: 1. there was lack of adequate lecturer development for the use of VR tools for teaching; 2. There was inadequate VR tools for teaching and learning; 3. the use of VR was a problem to implement since these resources were not tailor-made for the current curricula that the lecturers were expected to effectively deliver; 4. Lectures were not compatible with the use of VR tools as they lack adequate training; and 5. There was a lack of adequate funding for the procurement of 4IR resources.

It was evident that there were significant dynamics that prevent some lecturers from the use of VR tools to equip pre-service teachers. Lecturers need ongoing support to be able to use 4IR tools, in this case, VR resources for teaching, which are described as some of the most effective and necessary teaching and learning resources of the this revolution [11]. To add on to this view, it is also suggested that students retain more information and can better apply what they will have learned after participating in VR exercises [12].

Based on these findings, the study recommends that the departments in which the participants of this study belong to should urgently provide all the support required to facilitate the lecturers' use of VR resources for teaching. This should be done through engaging intensive professional development at all levels; and through the procurement of all relevant VR tools; as well realigning the current curricula to suit the use of VR resources. It is highly recommended that the lecturers and students in this institution urgently access the opportunities of VR resources as this would provide lifelike and collaborative teaching and learning environments. Moreover, the use of VR resources would enhance students' concrete experience and active experimentation with phenomena. Without VR resources, lecturers would have no choice but continue to use traditional pedagogy, which hinders the comprehension of abstract concepts, as there would be no room for concrete experiences and reflective observation by students.

It is therefore, of utmost importance that VR be employed within teaching and learning environments as it provides new forms and methods of visualization, drawing on the strengths of visual representations [27]. Furthermore, this researcher maintains that the use of VR provides an alternate method for presentation of material, and in some cases, it can more accurately elucidate some configurations and processes than by other means, allowing extreme close-up examination of an object, observation from a great distance, and observation and examination of areas and events unavailable by other means [27]. Hence, this study strongly recommends use of VR resources by lecturers and students in their teaching and learning environments in order to effectively prepare the novice teachers to infuse these resources in their practices as expected in this 4IR era.

Acknowledgements

We would like to recognize and give special thanks and gratitude to the university staff, particularly those who participated in the interviews. We would also like to thank all of the other people who have shown interest in our research project and have offered kind words of encouragement and optimism as we endeavored to improve pedagogy through the use of relevant fourth industrial revolution tools. We are eternally indebted to you and inspired by you beyond imagination.

Conflict of interest

We do hereby declare that there is no conflict of interest for the following reasons:

i. All forms of financial support are acknowledged in our contribution.

ii. There is no commercial or financial involvements that might present an appearance of a conflict of interest related to the Contribution.

iii. We have not signed an agreement with any sponsor of the research reported in the Contribution that prevents us from publishing both positive and negative results or that forbids us from publishing this research without the prior approval of the sponsor.

iv. We have checked the manuscript submission guidelines to see whether the journal requires a Declaration of Conflicting Interests and have complied with the requirements specified where such a policy exists.

Declarations

We hereby declare that this study titled, "Lecturers' experiences in teaching using VR resources at a selected university, South Africa" is our original work and it was never used or undertaken anywhere else. All the sources that the study employed have been mentioned and acknowledged by way of complete references. This study has been supported through a partnership between the South African Department of Higher Education and Training and the Tshwane University of Technology's Department of Research and Innovation.

Author details

Olika Moila[1*], Andile Mji[2] and Sibongile Simelane-Mnisi[3]

1 Faculty of Humanities, Department of Technology and Vocational Education, Tshwane University of Technology, Pretoria, South Africa

2 Faculty of Humanities, Office of the Executive Dean, Tshwane University of Technology, Pretoria, South Africa

3 Faculty of Science, Department of Curriculum Development and Support: eLearning, Tshwane University of Technology, Pretoria, South Africa

*Address all correspondence to: moilao@tut.ac.za

References

[1] Zawacki-Richter O., Latchem C. (2018). Exploring four decades of research in computers & education *Computers & Education,* 122, pp. 136-152

[2] Boulton C.A., Kent C., Williams H.T. (2018). Virtual learning environment engagement and learning outcomes at a 'bricks-and-mortar'university. *Computers & Education,* 126 pp. 129-142.

[3] Muñoz-Cristóbal J.A., Gallego-Lema V., Arribas-Cubero H.F., Martínez-Monés A., Asensio-Pérez J.I. (2017). Using virtual learning environments in bricolage mode for orchestrating learning situations across physical and virtual spaces. *Computers & Education*, 109, pp. 233-252.

[4] Wang R., Wiesemes R., Gibbons C. (2012). Developing digital fluency through ubiquitous mobile devices: Findings from a small-scale study. *Computers & Education*, 58 (1) pp. 570-578.

[5] Slater,M. & Sanchez-Vives, M.V. (2016). Enhancing our lives with immersive virtual reality, Front. Robot. AI. 3, 1-47, http://dx.doi.org/10.3389/frobt.2016.

[6] Dalgarno, B. & Lee, M.J.W. (2010). What are the learning affordances of 3-d virtual environments? Br. J. Educ. Technol. 41, 10-32, http://dx.doi.org/10.1111/j.1467-8535.2009.01038.x.

[7] Potkonjak, V., Gardner, M., Callaghan, V., Mattila, P., Guetl, C., Petrović, V. M., & Jovanović, K. (2016). Virtual laboratories for education in science, technology, and engineering: A review. Computers & Education, 95, 309-327.

[8] Freina, L., & Ott, M. (2015). A literature review on immersive virtual reality in education: state of the art and perspectives. In The international scientific conference eLearning and software for education (Vol. 1) (p. 133). "Carol I" National Defence University.

[9] Nooriafshar, R.; Mehryar, M.; & Williams, T. (2004). The use of virtual reality in education. In: American Society of Business and Behavioral Sciences (ASBBS) Seventh Annual International Conference, 6-8 Aug 2004, Cairns, Australia.

[10] Beier K.P. (2004). Virtual Reality: A Short Introduction. http://www-vrl.umich.edu/intro/

[11] Rogers, S. (2019). Virtual reality: The learning aid of the 21st century. *Forbes*.

[12] Krokos, E., Plaisant, C. & Varshney, A. (2019). Virtual memory palaces: immersion aids recall. Virtual Reality, 23(1), 1-15.

[13] Parton, G. (2018). Virtual Reality in Education: The VR Classroom. https://www.leedsbeckett.ac.uk/blogs/carnegie-education/2018/11/virtual-reality-in-education-the-vr-classroom/

[14] Burns, M. (2016). Immersive Learning for Teacher Professional Development, eCampus News: http://www.ecampusnews.com/technologies/

[15] Abrosimova, K. (2015). 5 ways virtual reality will change education. [Online]. Available: http://www.hypergridbusiness.com/2014/09/5-ways-virtual-reality-will-change-education/

[16] Jenner, M. (November 2015). 99p Virtual Reality and the implications for video in HE. [Online]. Available: https://blogs.ucl.ac.uk/ele/2015/11/16/99p-virtual-reality-and-the-implications-for-video-in-he/

[17] Ramirez, M. (2015). "Five emerging trends for innovative tech in education," JISC, September 2015.

[18] Häfner, P., Dücker, J., Schlatt, C., & Ovtcharova, J. (2018). Decision support methods for using virtual reality in education based on a cost-benefit-analyses. Paper presented at The 4th International Conference of the Virtual and Augmented Reality in Education (VARE 2018), September 17-18 2018, Budapest, Hungary.

[19] Hassan, A. (2021). Virtual Reality in Education: Achievements and Challenges. https://www.dynamologic.com/

[20] Kouri, E. (2019). Augmented & Virtual Reality. https://www.abiresearch.com/blogs/KT/

[21] Alfalah, S. (2018). Perceptions toward adopting virtual reality as a teaching aid in information technology. Education and Information Technologies, 23(6), 2633-2653. https://doi.org/10.1007/s10639-018-9734-2.

[22] Minocha, S., Tudor, A.-D., & Tilling, S. (2017). Affordances of mobile virtual reality and their role in learning and teaching. In L. Hall, T. Flint, S. O'Hara, & P. Turner (Eds.), Proceedings of the 31st international BCS human computer interaction conference (HCI 2017). 3-7 July, Swindon, UK. https://doi.org/10.14236/ewic/HCI2017.44.

[23] Krauss, S. (2005). Research Paradigm and Meaning Making: a Primer. Qualitative Report, 10, 4, 758-770.

[24] Marshall, M. (1996). Sampling for qualitative research. Family Practice, 13, 6, 522-525.

[25] McMillan, J.H. & Schumacher, S. (2010). Research in Education: Evidence-based Inquiry Edition. (7th Ed.). New Jersey: Pearson.

[26] Guba, E.G. and Lincoln, Y.S. (1994). Competing Paradigms in Qualitative Research. In: Denzin, N.K. and Lincoln, Y.S. (Eds), Handbook of Qualitative Research. Thousand Oaks, CA: Sage, 105-117.

[27] Pantelidis, V. S. (1995). Reasons to use virtual reality in education. VR in the Schools, 1(1), 9. Retrieved from http://vr.coe.ecu.edu/vrits/1-1pante.htm

Chapter 3

Cytopathology Using High Resolution Digital Holographic Microscopy

Sarita Ahlawat, Purnima Sharma, Ankita Pandey, Durga Bisht, Aanisa Jan, Apoorv Pant, Ritika Malik, Sandeep R. Mathur, Kiran Agarwal, Smita Singh, Meeta Singh and Kedar Khare

Abstract

We summarize a study involving simultaneous imaging of cervical cells from Pap-smear samples using bright-field and quantitative phase microscopy. The optimization approach to phase reconstruction used in our study enables full diffraction limited performance from single-shot holograms and is thus suitable for reducing cost of a quantitative phase microscope system. Over 48000 cervical cells from patient samples obtained from three clinical sites have been imaged in this study. The clinical sites used different sample preparation methodologies and the subjects represented a range of age groups and geographical diversity. Visual examination of quantitative phase images of cervical cell nuclei show distinct morphological features that we believe have not appeared in the prior literature. A PCA based analysis of numerical parameters derived from the bright-field and quantitative phase images of the cervical cells shows good separation of superficial, intermediate and abnormal cells. The distribution of phase based parameters of normal cells is also shown to be highly overlapping among different patients from the same clinical site, patients across different clinical sites and for two age groups (below and above 30 years), thus suggesting robustness and possibility of standardization of quantitative phase as an imaging modality for cell classification in future clinical usage.

Keywords: cervical cell imaging, quantitative phase, cell classification and characterization, early cancer diagnosis

1. Introduction

Cervical cancer is the fourth most prevalent cancer among women worldwide. Human Papillomavirus (HPV) is known to be the main cause of cervical cancer [1]. It is well-known that cervical cancer has long latency period. Pre-malignant abnormalities in cervical cells can take up to a decade to progress to carcinoma. Early diagnosis of pre-cancerous cervical cells and treatment help in halting the progression of this fatal cancer [2]. The five year survival rate for patients suffering from cervical cancer has been documented to be over 60% [3]. In spite of the long latency period for cervical cancer to develop, the mortality rate among cervical cancer patients is high in developing countries due to shortage of skilled clinicians and lack of effective screening tools [4, 5].

The cervix which is the innermost part of uterus is sub-divided between the endo-cervix and ecto-cervix regions. The endo-cervix is composed of glandular cells whereas the ecto-cervix is made up of squamous cells. Transformation zone is the place where these regions adjoin, and where most of the cervical cancer is known to originate [6]. Starting with the transformation zone the cells in the squamous region are typically classified as basal, para-basal, intermediate and superficial respectively. The majority of cells in a typical Pap-smear cell sample used for examination by clinicians are from the intermediate and superficial outer layers. The classification of precancerous cells as low grade and high grade is established through Bethesda system [7, 8] which is based on morphological changes in the cells (particularly the cell nuclei). Traditionally the detection of precancerous cervical cells is primarily performed using cytological screening. The widespread usage of liquid based cytology (LBC) in recent years has made the process of sample preparation and examination more uniform. However, screening methods based on visual inspection can suffer from both inter- and intra-observer variability. Machine learning based approaches have gained attention in this regard [9–13] and standard benchmark datasets of cervical cell images have been created [14, 15]. The goal of machine learning approaches is to make the process of cell classification at least semi-automated and to provide an assisting tool for cyto-pathologists.

The machine learning based studies have focused mainly on the 2D bright-field images of the cervical cells and their nuclei. Since cells are 3D objects, we believe that additional morphological information in the third (depth) dimension of cells, if available, can provide new information and help any image based cell classification. Digital Holographic Microscopy (DHM) is an interferometric imaging technology [16–18] which can fill this gap and provide quantitative phase information that may then be related to the depth dimension of the cells. When a coherent beam of light of wavelength λ is transmitted through a cell, the wave-front undergoes a phase change given by:

$$\phi(x,y) = \frac{2\pi}{\lambda}\int dz \;\; n(x,y,z). \tag{1}$$

Here n(x,y,z) represents the refractive index distribution within the cell relative to its surroundings. It is important to understand that phase provides new non-redundant information that cannot be derived from the usual 2D bright-field images. The phase change $\phi(x,y)$ cannot be measured directly by a 2D array sensor but may be recorded in the form of an interference fringe pattern. Here the coherent light source is first split into two beams, one of the beams passes through the cell sample and the other reference beam travels through free space before the two beams are recombined to record an interference pattern. As per Eq. (1), the phase function contains information on optical path length (product of refractive index and thickness) through the cell sample at location (x,y). While quantitative phase images have been shown to provide interesting new information about cancer cell morphology [19–22], clinically this modality is not yet popular and clinician are not trained to interpret quantitative phase images. We therefore follow a protocol where a focused bright-field image of a cervical cell is recorded along with phase image of the cell in the same focus plane. This way the clinicians can correlate with their traditional knowledge and treat phase images as an additional channel of morphological information. Recently we demonstrated such an approach for unsupervised organization of cervical cell images [23]. In the present imaging study over a much larger sample size with samples collected from different clinical sites, we examine the structural changes in phase images of cervical cell nuclei and highlight their potential importance for cell classification. Even though 2D images are able to distinguish between the major stages of normal cells, the phase images allow one to observe the morphological changes as the cells evolve through these stages. Additionally we examine the structural characteristics of abnormal samples as identified by practicing cyto-pathologists.

Traditional DHM systems are based on single-shot off-axis interference configuration or the multi-shot phase shifting con-figuration. The single-shot off-axis systems are simpler and cheaper to build but the conventional Fourier filtering approach for phase reconstruction in these systems leads to sub-optimal phase resolution. The multi-shot phase shifting configurations offer full resolution but are hardware intensive and require stringent vibration isolation thus making them difficult to employ in clinical settings. In recent years our group has developed optimization based phase reconstruction algorithms [24–27] for single-shot DHM systems which offer the simplicity of hardware without compromising on resolution and quantitative phase accuracy. The full diffraction-limited resolution capability of our system allows us to treat the bright-field and phase images on par (with respect to their lateral resolution). The single-shot operation also reduces the cost of building a DHM system making it more accessible for wider deployment. Based on our imaging study we find that the phase images contain important morphological information associated with different classes of normal as well as abnormal cells. Further this phase information is seen to be robust across the samples from three clinical sites. Also the samples consisted of age group of 17–60 years of the subjects. The cell morphology captured as numerical parameters from the phase images can provide valuable additional information to clinicians over what they usually access with routine bright-field microscopy. Our results suggest that phase imaging can become an important clinical modality, and it should be possible to design phase-based software tools for clinicians to make better informed decisions with this new information. The Chapter is organized as follows. In Section 2 we explain the technique of digital holographic microscopy (DHM) and the nature of quantitative phase images along with our phase reconstruction methodology. Section 3 briefly describes the details of the cell samples used. The results are discussed in Section 4. In Section 4.1 we start by showing images of cervical cells in both bright-field and phase modalities to illustrate morphological changes in cervical cell nuclei. This is followed by PCA analysis of the cell data based on the morphological parameters derived from the cell images in Section 4.2. In Section 4.3 we describe our analysis to understand if the most important phase parameters for normal cells are consistent across different patients from same clinical site, across different clinical sites and between different age groups. Finally in Section 5 we provide concluding remarks.

2. Digital holographic microscopy (DHM)

DHM is an interferometric modality where the recorded image data $H(x,y)$ represents interference between the object wave $O(x,y)$ representing the light which has interacted with the cell sample and the reference wave $R(x,y)$ that has not interacted with the sample is described as:

$$H = |R|^2 + |O|^2 + R^* O + RO^* . \qquad (2)$$

Here $*$ represents the complex conjugation of the corresponding wave-function. In the image plane holography case as in the present study, $O(x,y)$ represents the resultant image field corresponding to the cell sample slide when observed through a $40x$ infinity corrected imaging system [23]. The interference is possible due to the use of a laser source which ensures that the object and reference waves remain temporally coherent at the detector plane and produce interference fringes with good contrast. Since our DHM system is also fitted with a white light LED illumination which allows recording of the cell sample in the usual bright-field mode for ease of interpretation by a clinician.

Reconstruction of single-shot holograms is traditionally performed using the Fourier transform method. However, due to the low-pass filtering nature of this method image plane phase recoveries with full pixel resolution cannot be obtained using this approach. This poses a problem as the bright-field images available will then seem to have higher resolution even though both have been recorded using the same microscope objective. In order to have both bright-field and phase images with same diffraction-limited resolution, we reconstruct of the complex object wave $O(x,y)$ using a sparse optimization method that has been developed by our group in recent years. In particular, recovery of the complex image field $O(x,y)$ is posed as an optimization problem where we minimize a cost function of the form:

$$\begin{aligned} C(O,O^*) &= C_1 + C_2 \\ &= \left\| H - \left(|R|^2 + |O|^2 + R^*O + RO^* \right) \right\|^2 + \psi(O,O^*). \end{aligned} \tag{3}$$

Here $\| \dots \|^2$ denotes the squared L2-norm of the quantity inside. The reference beam $R(x,y)$ is estimated by a separate calibration step involving recording of a straight line interference fringe pattern without any sample followed by accurate estimation of carrier frequency to fractional fringe accuracy [28]. The first term of the cost function represents the least square data fit and the second term $\psi(O,O^*)$ is a suitable image domain constraint. We use the modified Huber penalty function as a constraint and use an adaptive alternating minimization scheme explained in detail elsewhere [23, 26] for recovering the complex object function $O(x,y)$ in the image plane. The modified Huber penalty is defined as:

$$\psi(O,O^*) = \sum_{k=all\ \ pixels} \left[\sqrt{1 + \frac{|\nabla O_k|^2}{\delta^2}} - 1 \right]. \tag{4}$$

The tuning parameter δ is made proportional to the median of the gradient magnitudes of the image solution in a given iteration. The Huber penalty acts like the edge preserving Total Variation penalty at pixels where the gradient magnitude $|\nabla O|$ is much larger than δ and acts like the smoothing quadratic penalty for pixels where the gradient magnitude is small compared to δ. Further the adaptive optimization strategy makes sure that the change in the solution due to error minimizing step is balanced by that due to Huber minimization step in every iteration. We point out that the optimization problem above involves real valued data (hologram H) whose solution is complex valued. The steepest descent directions evaluated in the algorithm need to be evaluated using Wirtinger derivatives with respect to O^*. In particular we note that the Wirtinger derivatives for the two terms of the cost function in Eq. (3) is given by:

$$\nabla_{O^*} C_1 = -2\left[H - |R+O|^2\right] \cdot (R+O), \tag{5}$$

and

$$\nabla_{O^*} C_2 = -\nabla \cdot \left[\frac{\nabla O}{\sqrt{1 + \frac{|\nabla O|^2}{\delta^2}}} \right]. \tag{6}$$

It is important to note that the optimization procedure operates fully in the image domain making it possible to employ it over a region of interest and thus allowing full resolution reconstruction in near real time. In our study, a 256×256 pixel ROI phase reconstruction requires few seconds (<25 iterations) in a MATLAB implementation on a desktop with 3.1 GHz processor. The data consistency error for the reconstructed solution is within 5% relative error. A user can therefore select a region of interest near a cell nucleus for object wave reconstruction. The resolution and noise advantage of this optimization procedure over traditional Fourier filtering approach has been shown in a series of publications [24–27, 29], as a result, we will not discuss this point here once again. However for completeness we summarize the advantages of the optimization method in comparison to the traditional methods for image plane hologram processing in **Table 1**.

The phase map $\phi(x,y)$ as in Eq. (1) is the argument of the recovered complex object field $O(x,y)$ and is given by arctangent of the ratio of imaginary and real parts:

$$\phi(x,y) = \arctan\left(\frac{Im[O(x,y)]}{Re[O(x,y)]}\right). \tag{7}$$

Since the arctangent function is defined only over the range $[-\pi, \pi]$ the phase map defined in Eq. (4) is wrapped. A 2D unwrapping procedure based on the transport of intensity equation (TIE) [30] has been employed in our work in order to associate physical meaning to the phase map in accordance to Eq. (1). The steps involved in imaging are summarized in supplementary (**Figure 1**). A Pap-smear sample is first imaged in both bright-field and holographic modalities using a dual mode digital holographic microscope (fabricated by Holmarc Opto-Mechatronics Pvt. Ltd., Kochi, India). The holographic (or interferometric) image is used further

Processing method	Single/Multi-shot	Resolution
Fourier filtering	Single-shot	Low resolution
Phase shifting	Multi-shot	Full diffraction-limited
Optimization	Single-shot	Full diffraction-limited

Table 1.
Summary of resolution performance of image plane digital holographic methods.

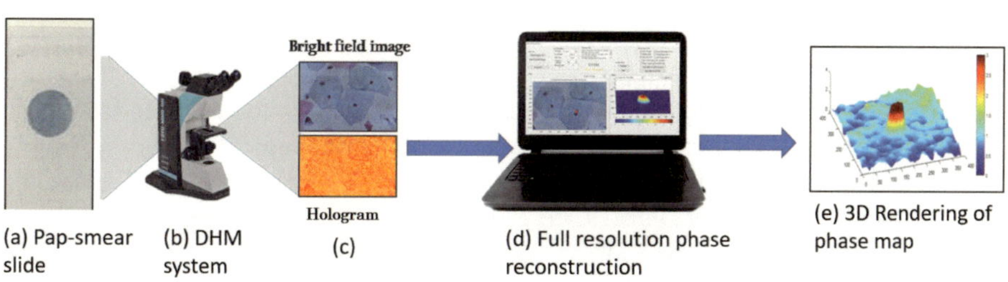

Figure 1.
Steps in imaging chain (a) Pap-smear slide, (b) dual-mode DHM system, (c) illustrative example of a bright-field image and a hologram recorded using the DHM system, (d) computer used for reading image from camera, phase reconstruction and computing morphological parameters from the bright-field and phase images, (e) illustrative example of phase map of a cell nucleus rendered as a surface plot.

Modality	Parameter	Parameter description[a]
Bright-field	Area	$\sum_{jk} M_{jk}$
	Nucleus perimeter	Number of boundary pixels of M
	Mean intensity of	R, G, B componentsAverage R, G, B values computed over mask M
	Variance of	R, G, B componentsVariance of R, G, B values computed over mask M.
	N/C Ratio	Ratio of areas (Nucleus)/(Cytoplasm) Labels assigned: low =1, mid =2, high = 3
Quantitative phase	Mean and maximum phase	$\overline{\phi}$ and maximum of ϕ computed over mask M
	Optical volume	(Area) × ($\overline{\phi}$) Computed over mask M.
	Variance of phase	Variance σ_ϕ^2 computed over mask M.
	Roughness at scales	1, 0.75, 0.5, 0.25 $\sum_{jk} \lvert(\nabla\phi)_{jk}\rvert$ computed over mask M.
	Moment of inertia	of nucleus $\sum_{jk} \left(\phi_{jk}\right) d_{jk}^2$ d_{jk} = distance between centroid and other pixels in mask M. Provides information about material distribution.
	Shift between geometric and phase centroid	$\lvert \vec{r}_{geom} - \vec{r}_{phase} \rvert$

[a]M denotes a binary (0, 1) mask for individual cell nucleus, q denotes phase map. Both are defined over ROI of 256 × 256 pixels centered on cell nucleus.

Table 2.
*Morphological parameters evaluated for each cell nucleus imaged in this study. More details about these parameters are provided in **Table 1** of ref. [23].*

for phase reconstruction as explained above. **Table 2** provides details about a number of morphological parameters derived from the cell images in the bright-field and quantitative phase modes. The morphological parameters were decided in consultation with practicing cyto-pathologists who participated in this study. We summarize them in **Table 2** for convenience of the reader. The N/C ratio which is the ratio of nucleus to cytoplasm areas has been included as list of three labels (low = 1, medium = 2, high = 3). This is because we found that a number of cells in the patient samples appeared in clusters and it was difficult to find boundaries of cytoplasm in simple automated manner in such cases.

3. Details of samples

We imaged a total of 48,006 cervical cells from 291 Pap-smear slides from three different hospitals in Delhi: AIIMS (All India Institute of Medical Sciences, New Delhi), LHMC (Lady Hardinge Medical College, New Delhi) and MAMC (Maulana Azad Medical College, New Delhi) (see **Table 3**). The samples were upto three years old (not from current patients) and stored in the repositories at the respective sites. They were collected by following the standard protocols for the Pap-smear examination. The patients varied in age from 16–70 years and came from varied geographical locations in India. The cell samples can be prepared conventionally or with Liquid-based cytology (LBC). For each method staining is performed with Pap stain for visualization with bright-field microscopy. For our study we have used

Clinical site	Preparation method	Number of patients normal / with condition [a]	Normal cells	Abnormal cells
AIIMS, New Delhi	Liquid based cytology	66 / 33	18801	532
	(ThinPrep)			
LHMC, New Delhi	Liquid based cytology	174 / 9	28007	214
	(SurePath)			
MAMC, New Delhi	Conventional	6/ 3	438	14
Totals		**246 / 45**	**47246**	**760**

[a] *As determined by clinicians.*

Table 3.
Details about cell samples used for imaging from the three clinical sites.

both types of samples. Two of the sites used liquid based cytology (LBC) slides prepared via ThinPrep and SurePath systems; while third site used conventional Pap-smears. In LBC method, samples are collected in liquid vials and the slide is prepared semi-automatically. The advantage of LBC is uniformity in sample preparation. On the other hand in conventional cytology the sample is applied directly to a slide for microscopic investigation. **Table 3** provides details about number of cells imaged. The classification of normal vs. abnormal cells in these samples was provided based on the bright-field images by practicing cyto-pathologists (S. R. M., M. S.,K. A., S.S.). The normal cells here include the superficial and intermediate cells while the abnormal cells include LSIL, HSIL, SCC, ASC-H and ASC-US type of cells [31].

4. Results

4.1 Illustrative bright-field and phase images of various cervical cell types

In this section we begin by providing sample images of cervical cell nuclei that were obtained using our dual-mode DHM system. While quantitative information obtained in terms of morphological parameters is certainly important, a large number of clinical sites worldwide typically use visual examination of cells using a bright-field microscope for cell classification. The importance of changes in nucleus structure in cancer diagnosis is already well-known [32]. With the illustrative examples in this section, we wish to qualitatively describe the morphological features observed in quantitative phase images of cervical cell nuclei in various stages. The simultaneous presentation of bright-field images (that pathologists can correlate to) and the quantitative phase images as shown here is important in our opinion from the perspective of clinical users. It is important to note that our single-shot full resolution phase reconstruction technique allows us to observe the quantitative phase images with the same resolution as the bright-field images. The examples shown here also aim to illustrate that the information contained in the quantitative phase images is different in nature from that in the bright-field images. Quantitative analysis of the images using morphological parameters as described in **Table 2** will be provided in the following sections. In **Figure 2** we show illustrative examples of normal cells in the intraepithelial squamous layer. A progression from intermediate to superficial stages is shown in **Figure 2(a)–(j)** respectively. As a cell progresses from intermediate to superficial stage the chromatin in the cell nucleus is known to condense. The superficial cells are in the outermost layer of ecto-cervix and have

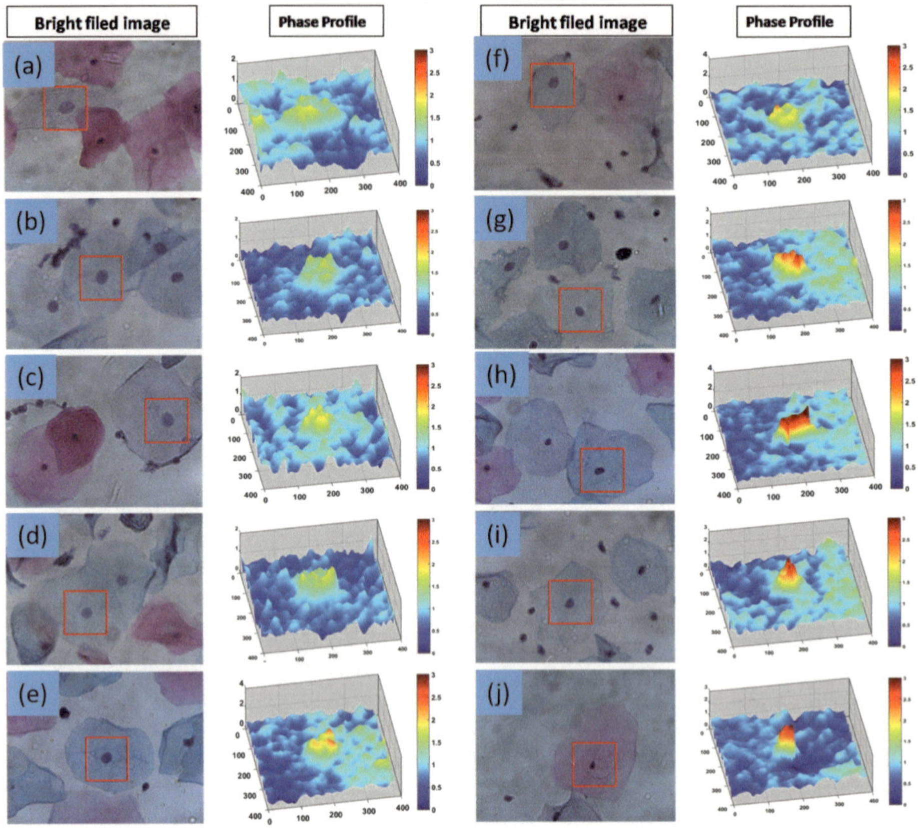

Figure 2.
Maturation of cells in intra-epithelial squamous layer from intermediate to superficial is shown from (a)-(j) in both bright-field and phase modes. The phase maps correspond to the ROIs marked in the bright-field images. As the nuclei progress to superficial stage, the nucleus area gets smaller and the phase profile is seen to get taller by approximately a factor of 2.

highly condensed pyknotic nucleus. While in the bright-field images the area of the nucleus is progressively decreasing and the color of the nucleus gets darker from intermediate to superficial stages, the accompanying phase map of the nucleus is seen to get taller by approximately a factor of 2. This change in the optical height profile cannot be inferred from the 2D bright-field images and the phase map is therefore seen to provide new morphological information. Further, from the phase profiles we also see that the evolution of the cells from intermediate to superficial stage happens via a continuous change. Next in **Figures 3–5** we examine the abnormal cell classes low grade squamous intraepithelial lesion (LSIL), high grade squamous intraepithelial lesion (HSIL) and squamous cell carcinoma (SCC) which progressively indicate higher grade abnormalities. The class LSIL consists of abnormal superficial and intermediate cells. Variable degrees of hyper-chromasia, nuclear size variation with coarsely granulated chromatin are identifiers of a typical LSIL cell. In the HSIL class, the degree of nuclear enlargement and hyper-chromasia is more than LSIL and the cells here are found in sheet-like aggregates. In both LSIL and HSIL cases the phase profiles of the cell nuclei are seen to have increased roughness or corrugations compared to the normal cells. In the SCC class which is considered to be a confirmed case of malignant cervical smear, the phase profile of the nucleus shows sharp narrow peaks with large phase values. It is once again important to note that the phase profile clearly provides new morphological information that is not readily available in the 2D bright-field images. Apart from the main classes above, the Bethesda system defines a class Atypical Squamous Cells

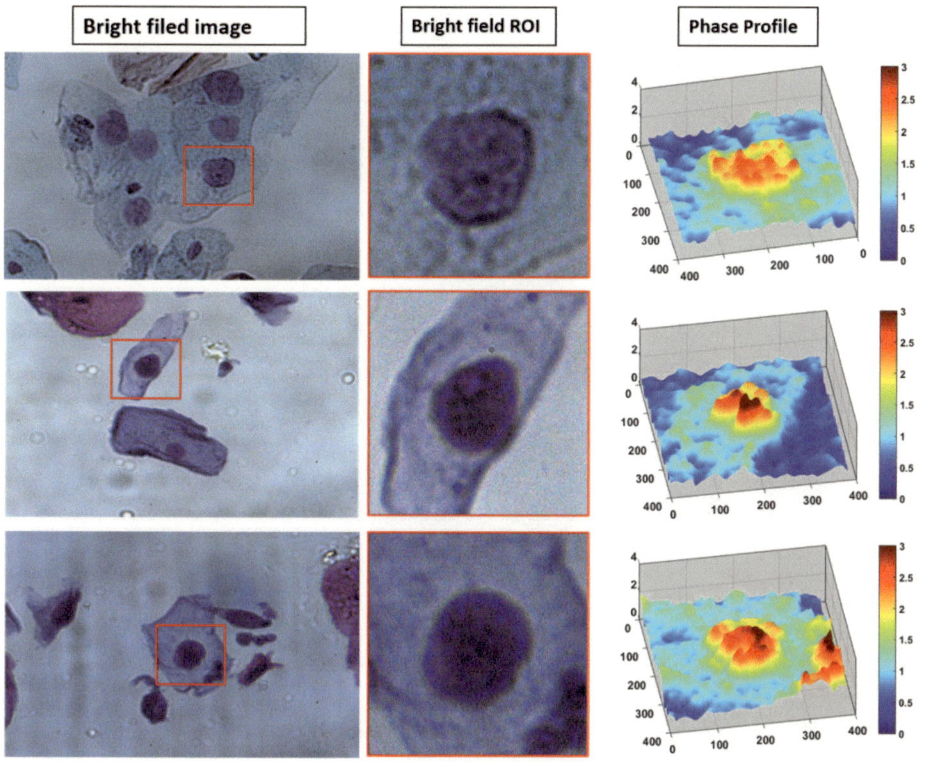

Figure 3.
Illustration of LSIL cells, the three columns show the bright-field image, selected ROI and the phase image of the ROI.

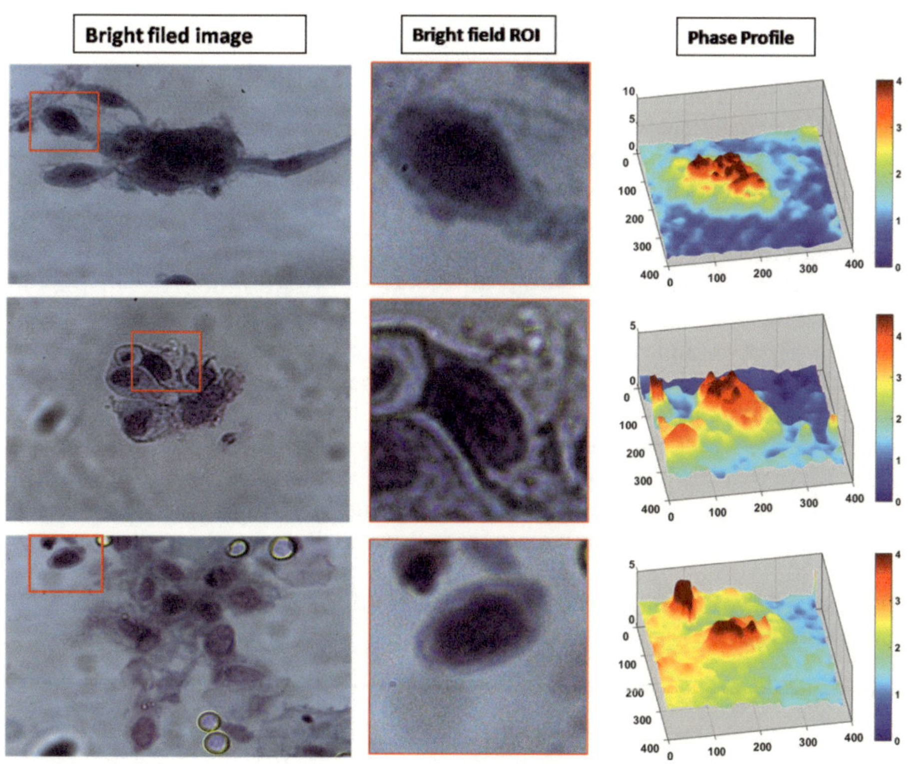

Figure 4.
Illustration of HSIL cells, the three columns show the bright-field image, selected ROI and the phase image of the ROI.

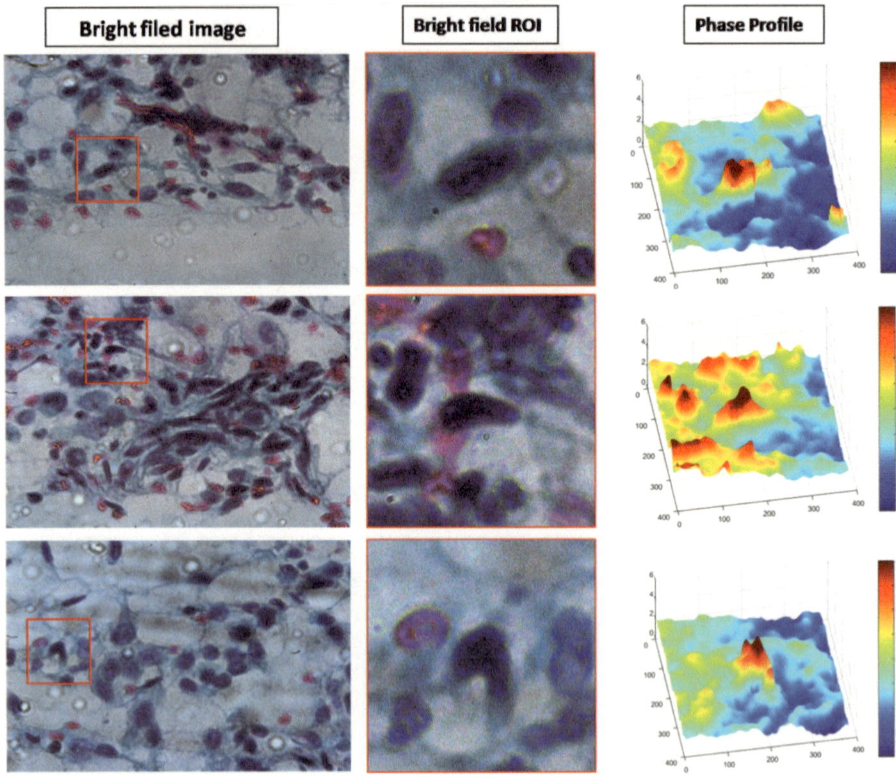

Figure 5.
Illustration of SCC cells, the three columns show the bright-field image, selected ROI and the phase image of the ROI.

ASC which include the samples that cannot be categorized as normal or abnormal typically. The nuclei in this class are typically larger in area, however, as seen in **Figure 6**, the average phase value in the nucleus is slightly lower compared to the LSIL, HSIL or SCC classes. In both ASC-US and ASC-H classes (**Figure 6**), the phase profile has undulations but the phase structure of ASC-H is flatter with lower average phase value in the nucleus as compared to that of the ASC-US cells. Further in some ASC nuclei, we observe local peaks in phase profile located near the nuclear boundary leading to a dip in the center. Finally we show rare abnormal cell cases in **Figure 7** that include inflamed, reactive, moon-type, virus-infected and koliocytotic classes. Just a few examples of these rare cells were present in our sample set. The phase profile for all these types appears corrugated with lower average phase values except for the virus infected cell type. While only a few representative images of each cell type have been shown simultaneously in bright-field and phase mode, we clearly observe that the phase profile offers distinctive morphological features that are not currently utilized in the clinical practice. This new information if incorporated in cyto-pathological examination, can be potentially valuable to clinicians.

4.2 PCA analysis of quantitative parameters obtained from bright-field and phase images of cervical cell nuclei

A MATLAB based software was designed to compute a number of morphological parameters associated with cell nuclei that are listed in **Table 2** from each of the cell nuclei imaged in bright-field as well as quantitative phase mode. The cell nucleus measurement data was therefore consisted of an ($N \times p$) matrix with N = 48,006

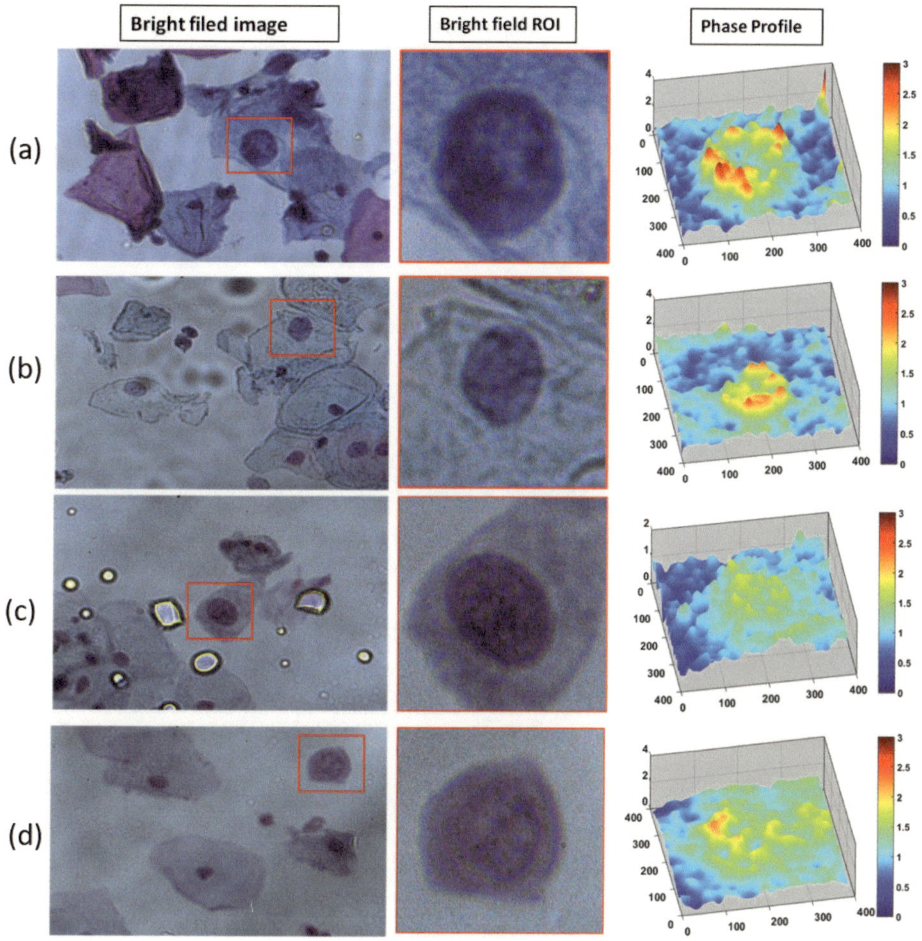

Figure 6.
Illustration of (a), (b): ASC-US cells and (c), (d): ASC-H cells, the three columns show the bright-field image, selected nucleus ROI and the phase image of the ROI.

and p = 20. Note that one parameter in the measurement set is the N/C ratio which was given three labels (low =1, mid =2, high =3) based on visual inspection of nuclei. This was done so that a number of cells of interest that appeared in clusters for which determining the cytoplasm boundary was difficult could be used in the analysis. All the other parameters were measured over the nucleus region in an automated fashion. For the present analysis, the cells were nominally labeled as superficial, intermediate and abnormal (including LSIL, HSIL, ASC-US, ASC-H, SCC) by practicing cyto-pathologists (S. R. M., M. S., K. A., S.S.). Since the number of abnormal cells was much smaller (1.6%) compared to the normal (superficial and intermediate) cells, a truncated data-set with 450 randomly selected cells from each of the three types (superficial, intermediate and abnormal) as per prior labelling was used to train the PCA. Denoting the truncated data matrix with 1350 rows and 20 columns (representing the measurements) with each column in standard form (zero mean and standard deviation 1) by A, the PCA solves the eigenvalue problem:

$$A^T A u_k = \mu_k u_k. \tag{8}$$

Here the superscript "T" stands for the transpose of the matrix A. The eigenvectors u_k are mutually orthogonal and are called as the principal vectors. All the

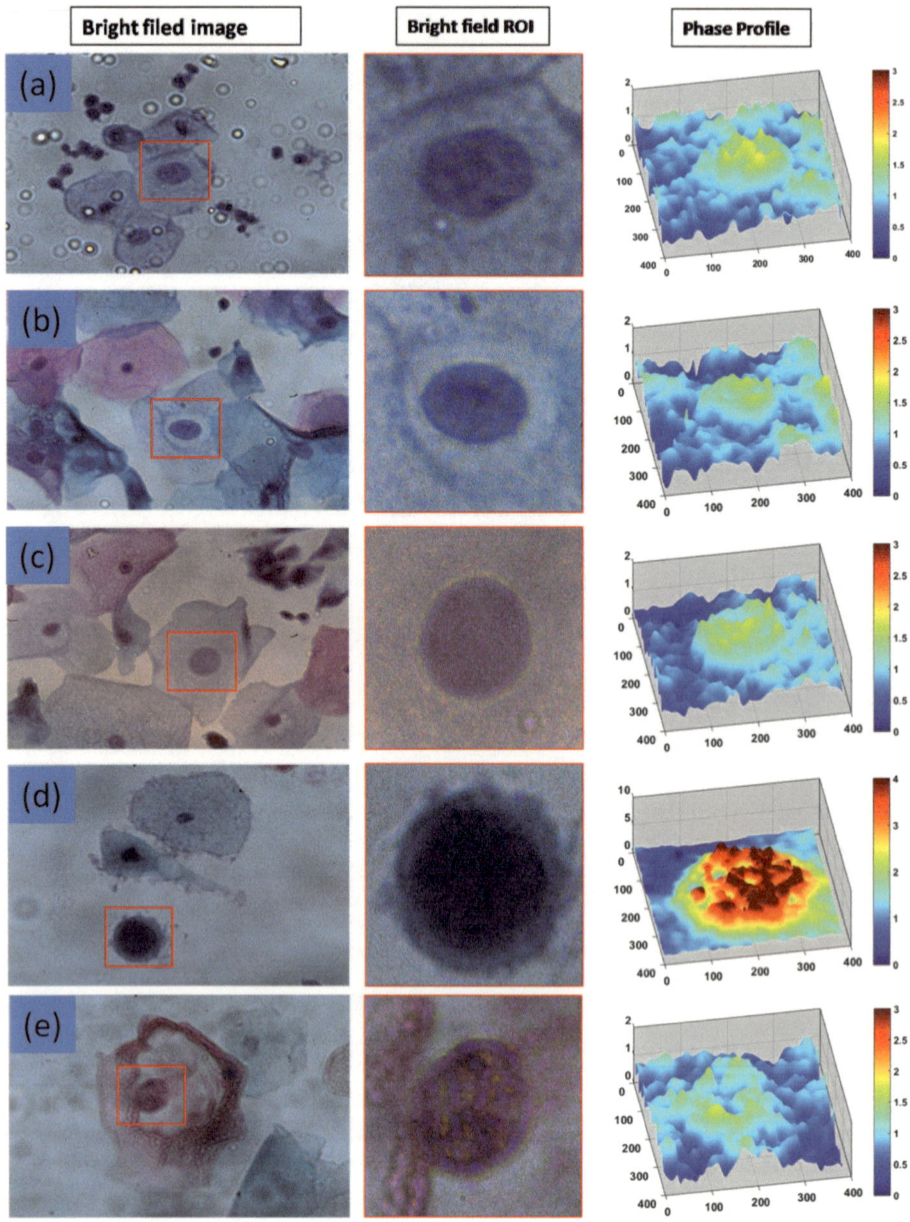

Figure 7.
Illustration of rare abnormal cell types: (a) inflamed, (b) reactive, (c) moon-type, (d) virus-infected and (e) koliocytotic; the three columns show the bright-field image, selected nucleus ROI and the phase image of the ROI.

cell data corresponding to the 48006 cells was then projected on the PCA vectors. The plot of first two components of PCA for all the cells is shown in **Figure 8**. The color coding of black, blue, and red corresponds to cells that were labeled separately by cyto-pathologists as superficial, intermediate and abnormal (all classes) respectively based on the bright-field images of the nuclei. Typical bright-field and phase images of cells from the three different regions of the PCA plot are also shown for illustration. The PCA plot based on bright-field and phase information separates most of the cells in three different classes, despite some overlap in adjacent classes. In particular, it is interesting to observe that almost all the cells labeled as abnormal fall in the bottom right corner of the PCA plot. We further examine

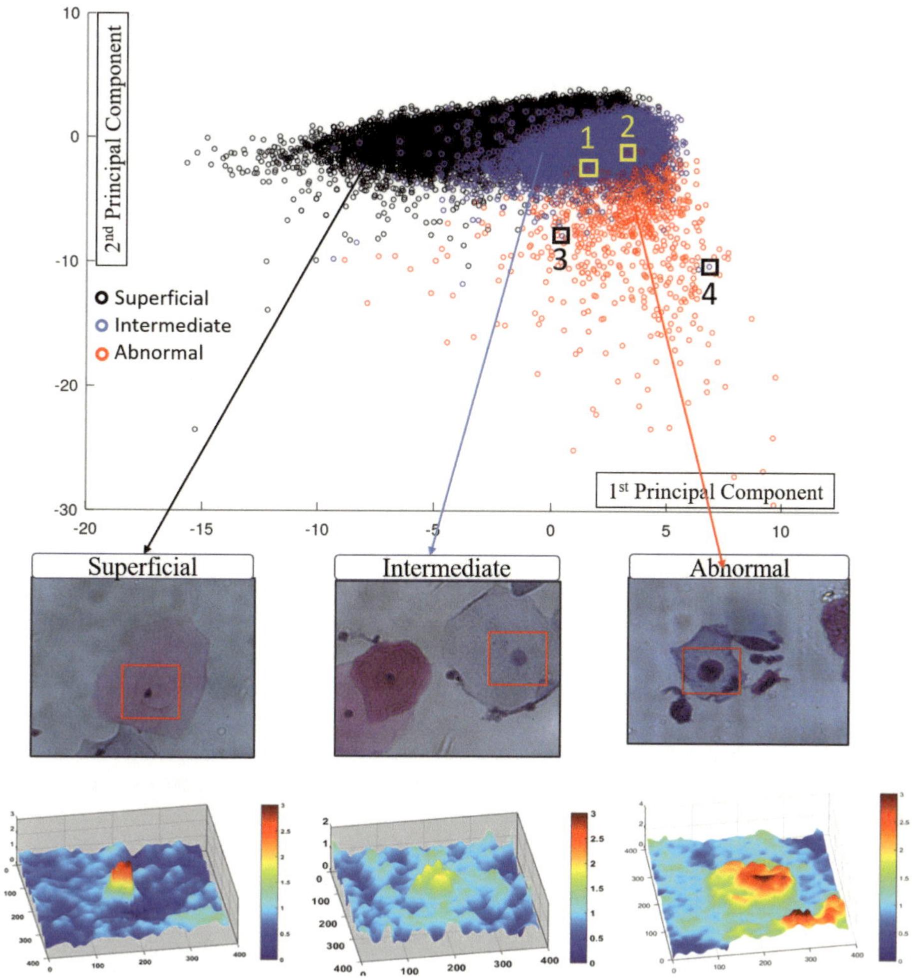

Figure 8.
Data corresponding to 48006 cells projected onto the first two PCA vectors. The color coding of black, blue, and red corresponds to cells that were labeled by cyto-pathologists as superficial, intermediate and abnormal (all classes) respectively.

four cells labeled as 1, 2, 3, 4 on the PCA plot that showed unexpected classification when phase parameters were used. Cells 1, 2 were labeled abnormal by the pathologist but were seen to be well within the intermediate region. Similarly the cells 3, 4 were labeled as intermediate but were observed to be well within the abnormal (red points) class. For a closer examination of this anomaly, we show bright-field and phase images of these cells in **Figure 9**. A re-examination of these cell images by pathologists suggested the following. Cell 1 is koliocytotic (abnormal) but it appears to have a dried up cytoplasm and leading to low phase values in the nucleus. The cell 2 is actually very similar to intermediate cells in general, but the pathologists labeled it as abnormal due to comparatively smaller sizes of other nuclei on the particular sample slide. The parameters associated with cell 3 are similar to abnormal cells but it is a rare example of enlarged intermediate cell. Finally cell 4 has folded cytoplasm leading to higher phase values although the cell may be considered to be of the intermediate class. Re-examination of these and other similar anomalies reveal that cervical cell classification has some aspects beyond simple numerical measurements performed on cell images (either in phase or bright-field modes) that need to be taken into account by any automated cell classification methodology. The issues like

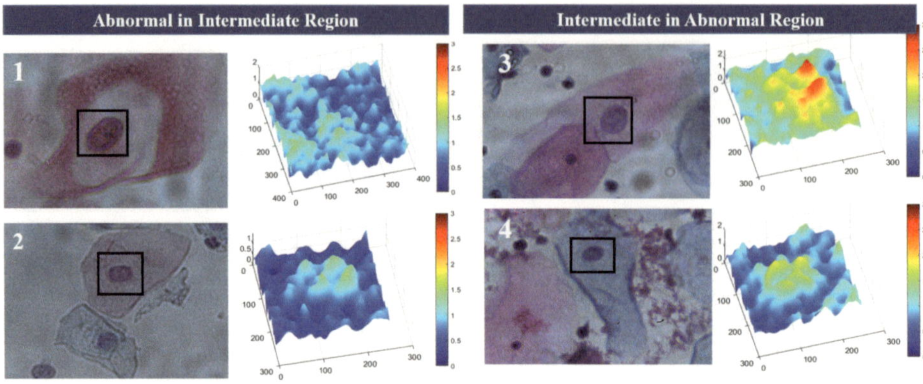

Figure 9.
*Examples of intermediate and abnormal cells falling well within the abnormal and intermediate regions of the PCA plot in **Figure 8**.*

folding of cell cytoplasm can for example be minimized with the LBC preparation methodology. PCA analysis was used here because the plot as in **Figure 8** can be generated essentially in an unsupervised manner, however, it is certainly not the best classification methodology available today. In future we hope to test the possibility of cell classification using more advanced machine learning ideas applied to this data.

We further performed a leave-one-out analysis of the PCA for the cell data to determine which of the 20 measurements influenced the PCA scores the most [33]. If the PCA eigenvectors are arranged as columns of a matrix U the scores Z for the data corresponding to the principal components may be expressed as:

$$Z = AU. \tag{9}$$

The plot in **Figure 8** thus corresponds to the first two columns of the score matrix Z. For the leave-one-out analysis, the PCA was performed at a time with only 19 parameters by leaving one of the measured parameters one by one. The data matrix, the eigenvector matrix and the score matrix corresponding to the case where j-th measurement ($j = 1, 2, 3, \dots, 20$) is left out may be denoted by $A^{(-j)}$, $U^{(-j)}$ and $Z^{(-j)}$ respectively. The importance of the j-th measurement is judged by the Procrustese distance D_j between the first $M = 2$ columns of the score matrices Z and $Z^{(-j)}$. A specific parameter will be judged to influence the PCA the most if its corresponding Procustes distance D_j is higher. The top five morphological parameters in order of importance are shown in **Table 4**. The relative Procrustes distances

Morphological parameter	Modality	Relative Procrustes Distance D_j
Moment of inertia	Phase	1.0
Optical volume	Phase	0.85
N/C Ratio	Bright-field	0.83
Perimeter of nucleus	Bright-field	0.79
Mean phase of nucleus	Phase	0.76

Table 4.
Relative importance of numerical parameters using leave-one-out analysis applied to PCA.

are calculated by dividing all the distances D_j with $(j = 1, 2, 3, \ldots, 20)$ by the maximum among them. We find that among the top five parameters that influenced the PCA scores the most, three were derived from the phase images while two were derived from the bright-field images. It may be noted from **Table 4** that two phase based parameters (moment of inertia and optical volume) influence the PCA more than the commonly used N/C ratio criterion. We therefore believe that quantitative phase may prove to be an important future imaging modality in addition to the commonly used bright-field microscopy for cervical cell classification.

4.3 Consistency of phase parameters

Quantitative phase is not a standard clinical methodology for cell classification, however, as we showed in Section 4.2, quantitative phase may become an important modality to consider for future clinical use. It is therefore important to understand if the phase parameters for cervical nuclei are consistent across different subjects from same clinical site, age group of subjects or clinical sites with different sample preparation methodologies. Since our leave-one-out PCA analysis suggested that optical volume and moment of inertia are the most important phase parameters as explained in the previous section, we have plotted a few hundred randomly selected normal cells (superficial and intermediate) with respect to these phase parameters in **Figure 10**. In **Figure 10(a)** we show the plot for 200 normal cells each for five different patients from a single clinical site. **Figure 10(b)** shows the same plot for 200 cells each from three different clinical sites with different sample preparation protocols. In **Figure 10(c)** we show the plot once again for 500 normal cells for two different age groups (below and above 30 years). From these plots we observe that the normal cells from different categories as above show highly overlapping distributions for the most important phase parameters. We believe that this observation

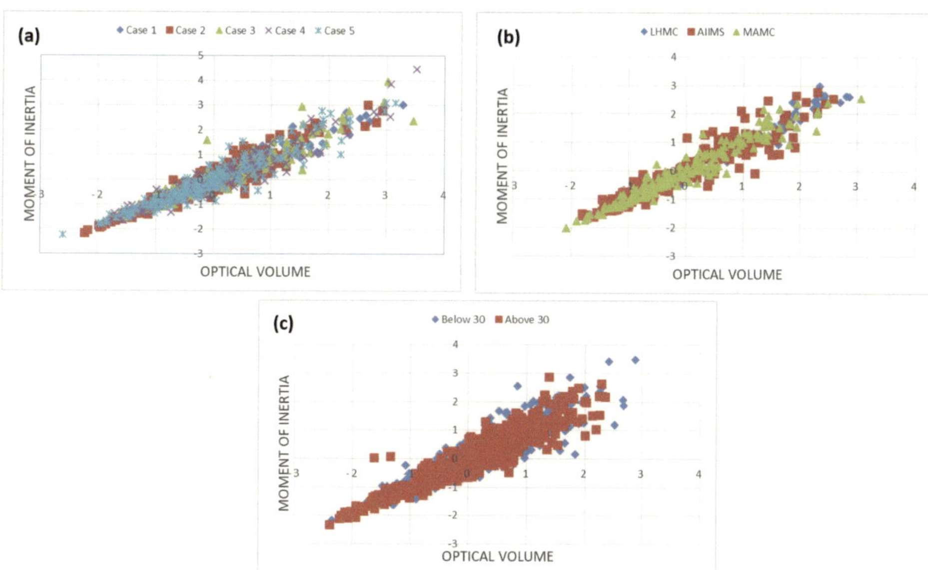

Figure 10.
Verification of consistency of the two most important phase parameters (moment of inertia and optical volume) decided based on the leave-one-out analysis; (a) plot of 200 cells each for 5 patients from the same clinical site (AIIMS), (b) plot of 200 cells each from three clinical sites with different sample preparation protocols, (c) plot of 500 cells each for patients below and above 30 years of age. The numerical values of moment of inertia and optical volume are normalized to standard form (zero mean and standard deviation one).

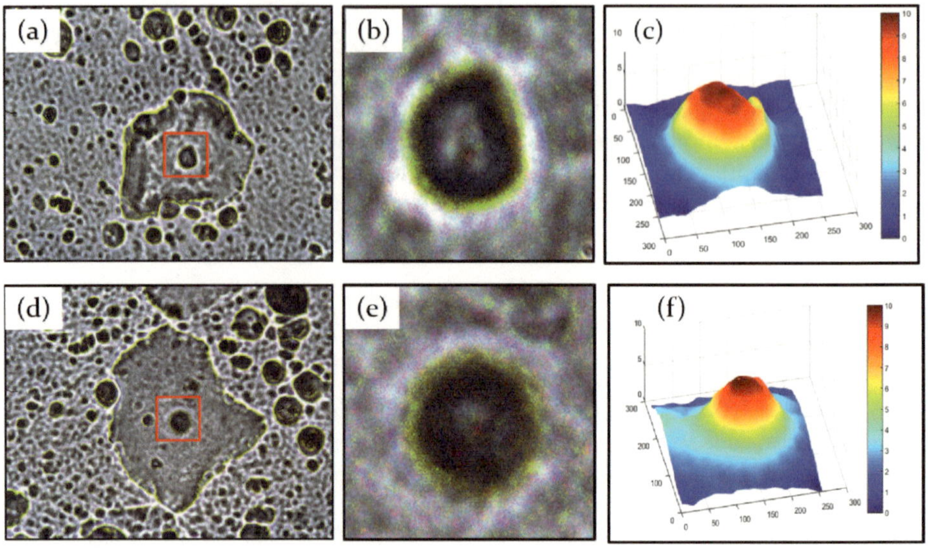

Figure 11.
Illustration of bright-field and phase imaging of unstained cervical cells: (a), (d): Bright-field images of unstained cells, (b), (e): Nucleus ROI selected from the bright-field image, (c), (f): Phase map of the unstained nuclei.

is very important for standardization and usage of quantitative phase imaging methodology in future clinical practice.

4.4 Observations on quantitative phase imaging of unstained cervical cells

In this section we briefly describe an interesting possibility of quantitative phase imaging of unstained cervical cell samples with two typical images of normal cells as shown in **Figure 11**. For unstained cervical cell samples, the Pap smear was prepared using the conventional method and cells were fixed with ethyl alcohol. While the staining protocols used in Pap-smear sample is a gold standard for diagnosis by cyto-pathologists, we note here that compared to stained cells, the phase signal observed from nuclei of unstained unprocessed cell samples is almost three times higher in magnitude. While interpretation of images of the unstained cells and their phase may require one to go through a learning process, the possibility of using unstained cell samples for diagnostic practice may offer an attractive alternative as the cell sample preparations will not involve any wet-lab processing and recurring costs associated with reagents.

5. Conclusions

In conclusion we have reported an image based study of cervical cells at various stages using bright-field as well as quantitative phase microscopy. Over 48000 cells have been imaged individually. The phase images of the cell nuclei were reconstructed using an optimization approach that provided same resolution as the bright-field images. This image data-set may be valuable for future application development using advance machine learning methods. The visual inspection of images shows interesting features in the phase images as the cells evolve from intermediate to superficial stages with distinct features associated with abnormal cells. This finding based on visual inspection is confirmed in the PCA analysis of the morphological parameters of cells derived from both the bright-field and phase images of cell nuclei. A leave-one-out analysis applied to the PCA scores suggests that apart from the N/C ratio that has been used for identifying abnormal cells for decades, the other two parameters that influence the PCA the most are optical volume and moment of inertia of nucleus - both of which are derived from phase images. A consistency study suggests that the phase parameters associated with normal cells show highly overlapping distributions for multiple patients from same clinical site, for three clinical sites with different sample preparation protocols and for patients in two age groups. The consistency of phase parameter distributions for these cases further suggest that phase is a robust modality that can certainly be used in a standardized manner in clinical practice. We believe that quantitative phase may become an important imaging modality in addition to the bright-field imaging that is solely used in the current clinical practice. While this study has been performed for cervical cells we believe that our conclusions regarding importance of quantitative phase may possibly have a wider applicability.

Acknowledgements

The authors acknowledge financial support from Department of Science and Technology India (Award: IDP/MED/34/2016) and Biotechnology Industry Research Assistance Council India (Awards: FT/12/02(110)/15/0237 and BT/SIBRI45/733/17). The unstained sample shown in **Figure 11** of this chapter was shared by Dr. R. Sankaranarayanan (RTI International India). Incubation support from FITT-IIT Delhi is gratefully acknowledged.

Conflict of interest

The authors declare no conflict of interest.

Author details

Sarita Ahlawat[1], Purnima Sharma[1], Ankita Pandey[1], Durga Bisht[1], Aanisa Jan[1], Apoorv Pant[2], Ritika Malik[2], Sandeep R. Mathur[3], Kiran Agarwal[4], Smita Singh[4], Meeta Singh[5] and Kedar Khare[2*]

1 Phase Laboratories Pvt Ltd, Technology Business Incubator Unit, Indian Institute of Technology Delhi, Hauz Khas, New Delhi, India

2 Department of Physics, Indian Institute of Technology Delhi, Hauz Khas, New Delhi, India

3 Department of Pathology, All India Institute of Medical Sciences, New Delhi, India

4 Department of Pathology, Lady Hardinge Medical College, New Delhi, India

5 Department of Pathology, Maulana Azad Medical College, New Delhi, India

*Address all correspondence to: kedark@physics.iitd.ac.in

References

[1] Schiffman M, Castle PE, Jeronimo J, Rodriguez AC, Wacholder S. Human papillomavirus and cervical cancer. The Lancet 2007; 370: 890–907.

[2] Andrae B, Andersson TML, Lambert PC, Kemetli L, Silfverdal L, Strander B, Ryd W, Dillner J, Tornberg S, Sparen P. Screening and cervical cancer cure: population based cohort study. BMJ 2012; 344:e900 doi: 10.1136/bmj.e900

[3] Benard VB, Watson M, Saraiya M, Harewood R, Townsend JS, Stroup AM, Weir HK, Allemani C. Cervical cancer survival in the United States by race and stage (2001–2009): Findings from the CONCORD-2 study. Cancer 2017; 123: 5119–5137.

[4] Cronje H. Screening for cervical cancer in developing countries. International Journal of Gynecology & Obstetrics 2004; 84: 101–108.

[5] Sankaranarayanan R, Nene BM, Shastri SS, Jayant K, Muwonge R, Budukh AM, Hingmire S, Malvi SG, Thorat R, Kothari A, Chinoy R, Kelkar R, Kane S, Desai S, Keskar VR, Rajeshwarkar R, Panse N, Dinshaw KA. HPV screening for cervical cancer in rural India. New England Journal of Medicine 2009; 360: 1385–1394.

[6] Small Jr W, Bacon MA, Bajaj A, Chuang LT, Fisher BJ, Harkenrider MM, Jhingran A, Kitchener H, Mileshkin LR, Viswanathan AN, Geffney DK. Cervical cancer: a global health crisis. Cancer 2017; 123(13): 2404–2412.

[7] Solomon D, Davey D, Kurman R, Moriarty A, O'Connor D, Prey M, Raab S, Sherman M, Wilbur D, Wright Jr. T, Young N. The 2001 Bethesda System: terminology for reporting results of cervical cytology. JAMA 2002; 287: 2114–2119.

[8] Nayar R, Wilbur DC (ed.s). The Bethesda system for reporting cervical cytology: definitions, criteria, and explanatory notes. Springer International Publishing 2015.

[9] Marinakis Y, Dounias G, Jantzen J. Pap smear diagnosis using a hybrid intelligent scheme focusing on genetic algorithm based feature selection and nearest neighbor classification. Computers in Biology and Medicine 2009; 39: 69–78.

[10] GençTav A, Aksoy S, ÖNder S. Unsupervised segmentation and classification of cervical cell images. Pattern Recognition 2012; 45: 4151–4168.

[11] Chankong T, Theera-Umpon N, Auephanwiriyakul S. Automatic cervical cell segmentation and classification in Pap smears. Computer Methods and Programs in Biomedicine 2014; 113: 539–556.

[12] Song Y, Zhang L, Chen S, Ni D, Lei B, Wang T. Accurate Segmentation of Cervical Cytoplasm and Nuclei Based on Multiscale Convolutional Network and Graph Partitioning. IEEE Trans on Biomedical Engineering 2015; 62:2421–2433.

[13] Zhang L, Le Lu, Nogues I, Summers RM, Liu S, Yao J. DeepPap: Deep Convolutional Networks for Cervical Cell Classification. IEEE Journal of Biomedical and Health Informatics 2017; 21: 1633–1643.

[14] Jantzen J, Norup J, Dounias G, Bjerregaard B. Pap-smear Benchmark Data For Pattern Classification. In: Proc. NiSIS (Nature Inspired Smart Information Systems); 2005: 1–9.

[15] Plissiti ME, Nikou C. Cervical cell classification based exclusively on nucleus features. In: Lecture Notes on

Computer Science Springer. 2012; 7325: 483–490.

[16] Kim MK. Principles and techniques of digital holographic microscopy. SPIE Reviews 2010; 1: 018005.

[17] Park Y, Depeursinge C, Popescu G. Quantitative phase imaging in biomedicine. Nature Photonics 2018; 12: 578–589.

[18] Kemper B, Illy E. Digital Holographic Microscopy. PhotonicsViews 2020; 17(1): 32–35.

[19] Mihailescu M, Paun I, Scarlat E, Grigorescu I, Nedelcu O, Radu R. Digital holographic microscopy for phase images of cervical cells 3D structure. In: Conference on Lasers and Electro-optics/Pacific Rim, Optical Society of America 2015: paper 27P69.

[20] Benzerdjeb N, Garbar C, Camparo P, Sevestre H. Digital holographic microscopy as screening tool for cervical cancer preliminary study. Cancer Cytopathology 2016; 124: 573–580.

[21] Lam VK, Nguyen TC, Chung BM, Nehmetallah G, Raub CB. Quantitative assessment of cancer cell morphology and motility using telecentric digital holographic microscopy and machine learning. Cytometry Part A 2018; 93: 334–345.

[22] Lam VK, Nguyen T, Phan T, Chung BM, Nehmetallah G, Raub CB. Machine Learning with Optical Phase Signatures for Phenotypic Profiling of Cell Lines. Cytometry Part A 2019; 95: 757–768.

[23] Mangal J, Monga R, Mathur SR, Ahlawat S, Khare K, Unsupervised organization of cervical cells using bright-field and single-shot digital holographic microscopy. Journal of Biophotonics 2019; 12: e201800409.

[24] Khare K, Samsheerali PT, Joseph J. Single shot high resolution digital holography. Opt. Express 2013; 21: 2581–2591.

[25] Singh M, Khare K. Single-shot interferogram analysis for accurate reconstruction of step phase objects. J. Opt. Soc. Am. A 2017; 34: 349–355.

[26] Singh M, Khare K. Single-shot full resolution region-of-interest (ROI) reconstruction in image plane digital holographic microscopy. J. Mod. Opt. 2018; 65: 1127–1134.

[27] Rajora S, Butola M, Khare K. Mean gradient descent: an optimization approach for single-shot interferogram analysis. J. Opt. Soc. Am. A 2019; 36: D7–D13.

[28] Singh M and Khare K. Accurate efficient carrier estimation for single shot digital holographic imaging. Opt. Lett. 2016; 41:4871–4874.

[29] Singh M, Khare K, Jha AK, Prabhakar S, Singh RP. Accurate multipixel phase measurement with classical-light interferometry. Phys. Rev. A 2015; 91:021801.

[30] Pandey N, Ghosh A, Khare K. Two-dimensional phase unwrapping using the transport of intensity equation. Applied Optics 2016; 55: 2418–2425.

[31] Gray W, McKee G. Diagnostic Cytopathology 2 ed. Churchill Livingstone, Philadelphia. 2002.

[32] Zink D, Fischer AH, Nickerson JA. Nuclear structure in cancer cells. Nature Reviews Cancer 2004; 4: 677–687.

[33] Krzanowski W. Cross-validation in principal component analysis. Biometrics 1987: 575–584.

Chapter 4

Latest Advances in Single and Multiwavelength Digital Holography and Holographic Microscopy

George Nehmetallah, Logan Williams and Thanh Nguyen

Abstract

In this Chapter, we discuss the latest advances in **d**igital **h**olography (DH) and **d**igital **h**olographic **m**icroscopy (DHM). Specifically, we study the different setup configurations such as single and multiwavelength approaches in reflection and transmission modes and the reconstruction algorithms used. We also propose two novel **t**elecentric recording configurations for single and **m**ulti-**w**avelength **d**igital **h**olographic **m**icroscopy (TMW-DHM) systems. Brief theory and results are shown for each of the experimental setups discussed. The advantages and disadvantages of the different configurations will be studied in details. Typical configuration features are, ease of phase reconstruction, speed, vertical measurement range without phase ambiguity, difficulty in applying optical and numerical post-processing aberration compensation methods. Aberrations can be due to: (a) misalignment, (b) multiwavelength method resulting in Chromatic aberrations, (c) the MO resulting in parabolic phase curvature, (d) the angle of the reference beam resulting in linear phase distortions, and (e) different optical components used in the setup, such as spherical aberration, astigmatism, coma, and distortion. We conclude that telecentric configuration eliminates the need of extensive digital automatic aberration compensation or the need for a second hologram's phase to be used to obtain the object phase map through subtraction. We also conclude that without a telecentric setup and even with post-processing a residual phase remains to perturb the measurement. Finally, a custom developed user-friendly graphical user interface (GUI) software is employed to automate the reconstruction processes for all configurations.

Keywords: digital holography, multi-wavelength digital holography

1. Introduction to digital holography (DH)

Digital holograms are generated by recording the interference pattern of two mutually coherent beams. These two beams are the object beam and the reference beam and the recording medium is usually a CCD [1]. The digital hologram recorded on the CCD due to the interference of the object beam E_O and the reference beam E_R is given by

$$h(x,y) \propto I_H = |E_R|^2 + |E_O|^2 + E_R^* E_O + E_O^* E_R, \tag{1}$$

where the * notation denotes the complex conjugate. Traditionally, in analog holography the reconstruction is performed by illuminating a holographic film by the conjugate of the reference beam E_R^*, and the real image is obtained from the last term of Eq. (1): $|E_R|^2 E_O^*$. The first two terms on the right hand side and the third term contribute to the zero order and the virtual image, respectively. The digital reconstruction is generally performed by numerically propagating the field $E_R^* h(x,y)$ by the recording distance, d or $-d$, to reconstruct either the real or virtual images. A typical schematic of the recording and reconstruction of DHs is shown in **Figure 1**. Several numerical reconstruction algorithms have been developed for DH, although the most common are the discrete Fresnel transform, the convolution approach, and reconstruction by angular spectrum. Each of these reconstruction algorithms will be subsequently briefly described.

1.1 Numerical reconstruction by discrete Fresnel transformation

The Fresnel Transform is based on the Fresnel approximation to the Huygens-Fresnel diffraction integral, and under the paraxial approximation, i.e., $d^3 >> (2\pi/\lambda)\left[(\xi - x)^2 + (\eta - y)^2\right]$, the reconstruction of the hologram can be approximated by the Fresnel transformation [1–6]:

$$\Gamma(\xi,\eta) = z(\xi,\eta)\mathfrak{F}_{x,y}\left[h(x,y)E_R^*(x,y)w(x,y)\right]\big|_{k_x=2\pi\xi/\lambda d, k_y=2\pi\eta/\lambda d}, \tag{2}$$

$$w(x,y) = \exp\left[-j\frac{\pi}{\lambda d}(x^2+y^2)\right], \tag{3}$$

$$z(\xi,\eta) = \frac{j}{\lambda d}\exp\left(-j\frac{2\pi d}{\lambda}\right)\exp\left[-j\pi(\xi^2+\eta^2)/\lambda d\right], \tag{4}$$

where $\mathfrak{F}_{x,y}\{\bullet\}$ is the Fourier transform operator. The intensity is calculated by squaring the optical field, i.e., $I(\xi,\eta) = |\Gamma(\xi,\eta)|^2$ and the phase is calculated using $\varphi(\xi,\eta) = \arctan\left(\mathrm{Im}[\Gamma(\xi,\eta)]/\mathrm{Re}[\Gamma(\xi,\eta)]\right)$. Since x,y are discretized on a CCD rectangular raster of N_x, N_y pixels of sizes Δx, Δy, the reconstructed image resolution in the ξ,η coordinates are given by [5–7].

$$\Delta\xi = \lambda d/N_x\Delta x, \Delta\eta = \lambda d/N_y\Delta y. \tag{5}$$

The image resolution given by Eq. (3) is considered to be "naturally scaled," such that the value of $\Delta\xi$ is automatically equal to the physical resolution limit imposed by the CCD sampled signal bandwidth [2, 6].

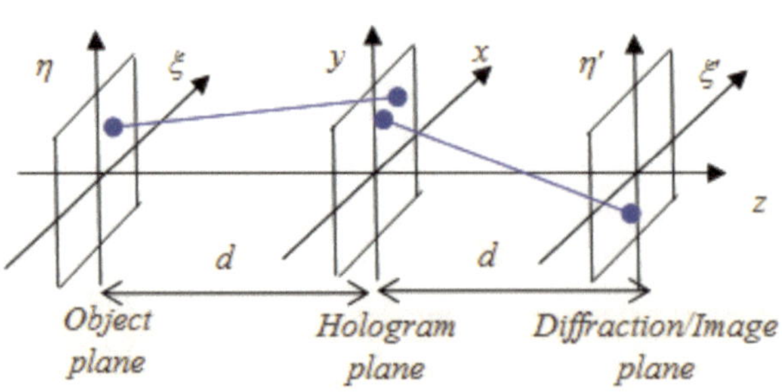

Figure 1.
Coordinate system for DH recording and reconstruction.

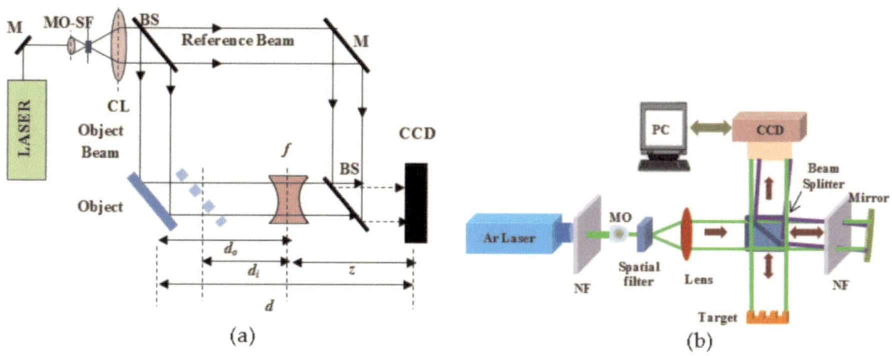

Figure 2.
Schematic of a DH setup (a) Mach-Zehnder setup, (b) Michelson setup. MO-SF: microscope objective-spatial filter, BS: beam splitter, NF: neutral density filter, M: mirror, CL: collimating Lens.

Figure 3.
The recorded hologram in (a) is reconstructed via Eq. (2) to yield (b) the reconstructed image. Note that (b) contains an in-focus (virtual) image on the right, and an out-of-focus (real) image on the left. The relevant reconstruction parameters are d = *39 cm,* λ = *496.5 nm,* Δx = *6.7 μm,* N = *1024, with reconstructed image resolution* $\Delta\xi = 28.5\ \mu m$.

A reflection type Fresnel DH setup based upon the Mach-Zehnder interferometer is schematically shown in **Figure 2(a)**. Light from a Laser source is divided into two parts with a beam splitter. One of the beams forms the reference, while the other is reflected off the object, then both interfere on a CCD camera to form a Fresnel hologram. **Figure 2(b)** shows a Michelson type setup [8, 9]. An example of a recorded hologram of a Newport Logo recorded using an Argon laser @ 496.5 nm and its reconstruction using Fresnel transform method are shown in **Figure 3(a)** and **(b)**, respectively.

1.2 Numerical reconstruction by the convolution approach

Since the diffracted field at a distance $z = d$ from the hologram can be expressed as

$$\Gamma(\xi,\eta) = \int_{-\infty}^{\infty}\int_{-\infty}^{\infty} h(x,y)E_R^*(x,y)g_{PSF}(x-\xi,y-\eta)\,dxdy, \tag{6}$$

the convolution approach can be written as:

$$\Gamma(\xi,\eta) = \left[h(\xi,\eta)E_R^*(\xi,\eta)\right] * g_{PSF}(\xi,\eta),\ g_{PSF}(\xi,\eta) = \frac{j}{\lambda}\frac{\exp\left(-jk_0\sqrt{d^2+\xi^2+\eta^2}\right)}{\sqrt{d^2+\xi^2+\eta^2}} \tag{7}$$

where the $*$ denotes convolution. Eq. (5) can be written as

$$\Gamma(\xi,\eta) = \Im_{x,y}^{-1}\left\{\Im_{x,y}\left(h\cdot E_R^*\right)\cdot\Im_{x,y}\left(g_{PSF}\right)\right\} \equiv \Im_{x,y}^{-1}\left\{\Im_{x,y}\left(h\cdot E_R^*\right)\cdot\left(G_{PSF}\right)\right\}, \tag{8}$$

where $G_{PSF} = \Im_{x,y}\left(g_{PSF}\right)$. Although the pixel sizes of the images reconstructed by the convolution approach are equal to that of the hologram, namely, $\Delta\xi = \Delta x, \Delta\eta = \Delta y$, the physical image resolution remains according to Eq. (3) and is ultimately governed by physical diffraction [5–7].

1.3 Numerical reconstruction by the angular spectrum approach

In Fourier space and across any plane the various spatial Fourier components of the complex field distribution of a monochromatic wave can be considered as plane waves traveling in different directions away from that plane. The field amplitude at any other point can be calculated by adding the weighted contributions of these plane waves, taking into account of the phase shifts they have undergone during propagation [1]. Similar to the convolution approach above, the angular spectrum approach is based on direct application of the propagation of the angular spectrum

Technique	Advantages	Disadvantages
Fresnel	• Fast (uses one FFT) • Used primarily for long distances. (Short distances possible with hologram upsampling prior to reconstruction.) • May be used for larger objects. • Image resolution can be arbitrarily scaled by applying zero padding or upscampling to the hologram.	• Image pixel size depends on reconstruction distance and wavelength. • Poor depth resolution for isolating adjacent hologram planes along the propagation axis (compared to newer methods, e.g. compressive sensing). • Not useful in inline holograms of scattering particles that have to be evaluated on different depths
Convolution	• Limited numerical image magnification is possible during reconstruction. • Image pixel size does not depend on distance and wavelength and is equal to hologram pixel size (Physical resolution still is governed by diffraction limit) • Useful in inline holograms of scattering particles that have to be evaluated on different depths	• Slower (Uses at least two FFT) • Used for small objects • Used for short distances. • Numerical image magnification does not improve object "resolution."
Angular spectrum	• Image pixel size is typically equal to the hologram pixel size. (Physical resolution still is governed by diffraction limit.) • May be used for very short distances where Fresnel technique fails (No minimum distance required between object and CCD).	• Typically used for in-line holograms • Slower (Uses at least two FFT) • Used for smaller objects that do not exceed the lateral extent of the CCD for in-line. • Zero padding techniques do not alter resolution.

Table 1.
Advantages and disadvantages of several digital holography reconstruction techniques.

of the field in the hologram plane. Accordingly, we define the angular spectrum of the field $(h \cdot E_R^*)$ at the hologram plane as [1]:

$$\tilde{E}_h(k_\xi, k_\eta) = \mathfrak{I}_{x,y}(h \cdot E_R^*) = \frac{1}{4\pi^2} \int_{-\infty}^{\infty} \int_{-\infty}^{\infty} (h \cdot E_R^*) \exp\left[j(k_\xi \xi + k_\eta \eta)\right] d\xi d\eta \tag{9}$$

where k_ξ, k_η are spatial frequency variables corresponding to ξ, η. After propagating a distance z, each plane wave component of the angular spectrum acquires an additional phase factor $e^{-jk_z z}$ where

$$k_z = \sqrt{k_0^2 - k_\xi^2 - k_\eta^2}. \tag{10}$$

Therefore, the reconstructed field at a distance $z = d$ becomes:

$$\begin{aligned} \Gamma(\xi, \eta) = \frac{1}{4\pi^2} \int_{-\infty}^{\infty} \int_{-\infty}^{\infty} \tilde{E}_h(k_\xi, k_\eta) \exp\left[-jd\sqrt{k_0{}^2 - k_\xi{}^2 - k_\eta{}^2}\right] \\ \times \exp\left[-j(k_\xi \xi + k_\eta \eta)\right] dk_\xi dk_\eta \end{aligned} \tag{11}$$

which is similar to Eq. (5) above.

Table 1 shows the advantages and disadvantages of the different reconstruction techniques discussed in this section.

2. Digital holographic microscopy (DHM)

DHM is usually applied to determine 3D shapes of small objects, with height excursions on the order of microns (or phase excursions on the order of a few radians). Since small objects are involved, a microscope objective (MO) is often used to zoom onto a small area of the object to enhance the transverse resolution. Holograms of microscopic objects recorded with DHM setups can be numerically reconstructed in amplitude and phase using the same DH reconstruction techniques discussed in Section 1. The phase aberrations due to the MO and the tilt from the reference beam have to be corrected to obtain the topographic profile or the phase map of the object [10–12]. **Figure 4(a)** and **(b)** show a Michelson DHM in reflection and transmission configurations, respectively.

For a reflective object on a reflective surface, the height profile on the sample surface is simply proportional to the reconstructed phase distribution $\varphi(\xi, \eta)$, through [10]:

$$h_z(\xi, \eta) = \left(\frac{\lambda}{4\pi}\right) \varphi(\xi, \eta). \tag{12}$$

For a transmissive phase object on a reflective surface, its thickness can be calculated as:

$$h_z(\xi, \eta) = \left(\frac{\lambda}{4\pi}\right) \frac{\varphi(\xi, \eta)}{\Delta n}, \tag{13}$$

where Δn is the difference of the index of refraction between the transparent object material and the surrounding medium (e.g. air).

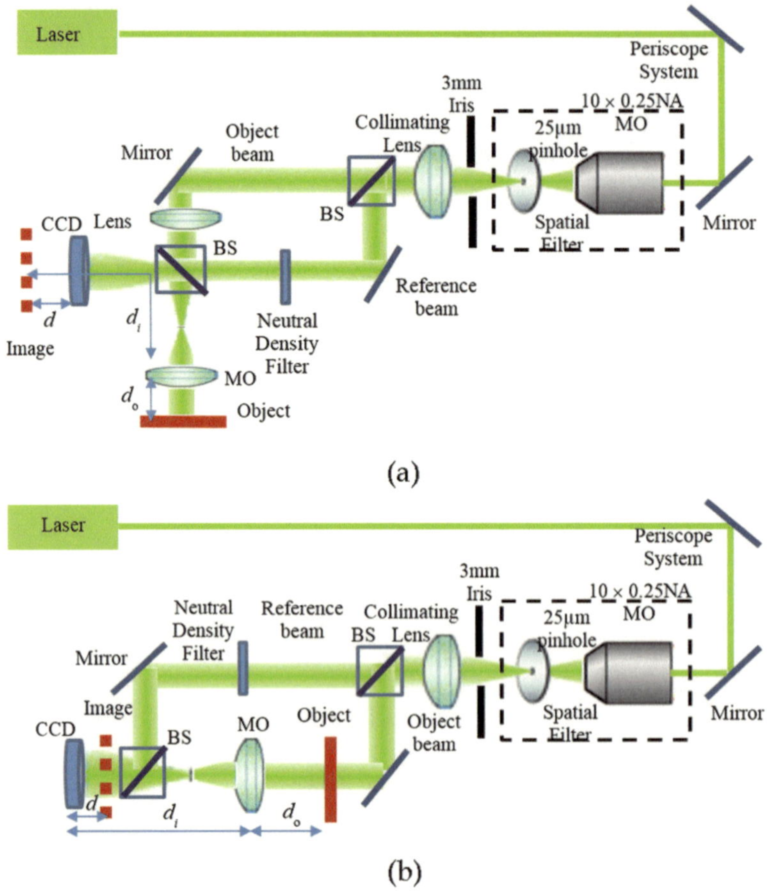

Figure 4.
Digital holographic microscope: (a) reflective setup, (b) transmissive setup.

For a transmissive phase object on a transmissive surface or between transmissive surfaces, the phase change (optical thickness) can be calculated as:

$$h_z(\xi,\eta) = \left(\frac{\lambda}{2\pi}\right)\frac{\varphi(\xi,\eta)}{\Delta n}. \tag{14}$$

As stated above, in DHM we introduce a MO to increase the spatial resolution which was computed according to Eq. (3). Due to the magnification 'M' introduced by the MO the pixel size in the image plane, $\Delta\xi_{mag}$ scales according to:

$$\Delta\xi_{mag} = \frac{\Delta\xi}{M} = \frac{\lambda d}{N.\Delta x \cdot M}, \tag{15}$$

which is simply the magnification predicted by geometric imaging. This is intuitively understood by realizing that the holographic recording is now simply a recording of the geometrically magnified virtual image located at distance d as shown in **Figure 5**. Thus, the pixel resolution is automatically scaled accordingly. We can enhance the transverse resolution approximately to be equal to the diffraction limit $0.61\lambda/\mathrm{N.A}$ of the MO, where N.A is the numerical aperture of the MO.

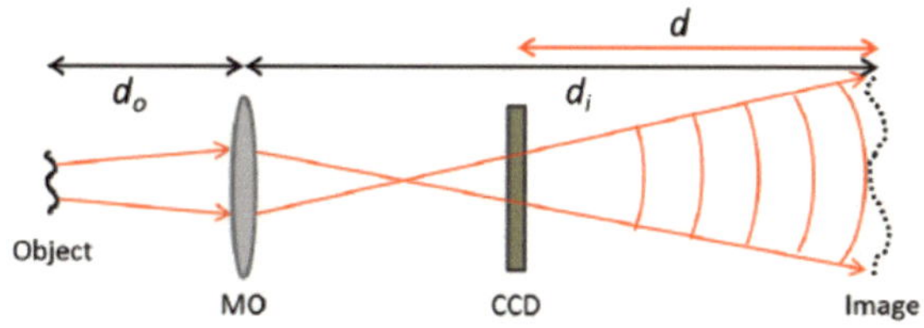

Figure 5.
Generalized holographic recording geometry using a lens. The image location is governed by geometric optics and may be on either side of the lens.

The complete reconstruction algorithm is governed by the equation [9–14]:

$$\Gamma(m,n) = \underbrace{Ae^{-j\left[\frac{\pi}{\lambda D}\left(m^2\Delta\xi^2+n^2\Delta\eta^2\right)\right]}}_{Quadratic\ \ Phase\ due\ to\ MO} z(m,n) \times \Im_{x,y}\left\{\underbrace{A_R e^{-j\frac{2\pi}{\lambda}\left(\sin\theta_x k\Delta x+\sin\theta_y l\Delta y\right)}}_{E^*_{\mathrm{Re}}} h(k,l)w(k,l)\right\}_{m,n}, \tag{16}$$

where $z(m,n) = \frac{j}{\lambda d}\exp\left(-j\frac{2\pi d}{\lambda}\right)\exp\left[-j\pi\lambda d\left(\frac{m^2}{N_x{}^2\Delta x^2}+\frac{n^2}{N_y{}^2\Delta y^2}\right)\right]$, $w(k,l) = \exp\left[-j\frac{\pi}{\lambda d}\left(k^2\Delta x^2+l^2\Delta y^2\right)\right]$ $\frac{1}{D} = \frac{1}{d_i}\left(1+\frac{d_o}{d_i}\right)$, and the focal length of the MO is: $\frac{1}{f} = \frac{1}{d_i}+\frac{1}{d_o}$.

Aberration compensation can be performed manually using a phase mask Ψ to cancel the effects of the quadratic phase due to the MO and the linear phase due to the reference tilt. The phase mask can be written as [11].

$$\Psi = A\exp\left\{j\left[\frac{2\pi}{\lambda}\left(m\Delta x\sin\theta_x+n\Delta y\sin\theta_y\right)\right]\right\}\times\exp\left\{j\left[\frac{\pi}{\lambda D}\left(m^2\Delta x^2+n^2\Delta y^2\right)\right]\right\}, \tag{17}$$

where θ_x, θ_y are the tilt angles of the reference beam and D is defined in Eq. (11). A more robust technique is to perform automatic aberration cancelation by approximating the residual phase front due to aberration using Zernike polynomials as explained in details in Refs. [13, 14], and shown in Section 4 below.

Consider the transmission setup shown in **Figure 4(b)**. A USAF 1951 resolution chart target is used as an object. The resolution of a USAF resolution chart is documented as:

$$R_{[lp/mm]} = 2^{\left[G+\frac{(E-1)}{6}\right]}, \tag{18}$$

where $R_{[lp/mm]}$ is resolution in line pair per millimeter, G is the group number, and E is the element number. (See **Figure 6(a)**). As an example, Group 4, Elements 3 and 4 has a resolution of 20.16 *lp/mm*, and 22.62 *lp/mm*, respectively. The wavelength used is λ = 488 nm, the reconstruction distance d = 0.202 m, D = 0.14 m, the magnification is $M \approx 8.25$, $k_{x0}/k_0 = \sin\theta_x = 0.01307$, $k_{y0}/k_0 = \sin\theta_y = 0.01305$. It should be noted that in practice, it is very difficult to obtain such precise parameter measurements in the laboratory. Typically, approximate measurements are made, then, varied slightly during numerical reconstruction to yield the "best focus" image. Such a process was followed to obtain the parameters listed above. In Section 4, we discuss the telecentric setups which mitigate these difficulties. **Figure 6(b)**

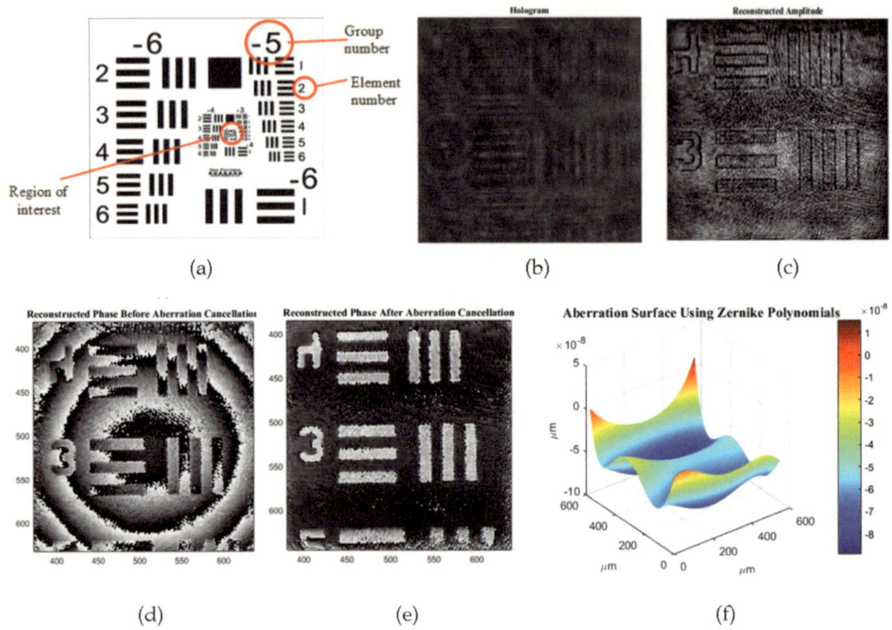

Figure 6.
(a) Schematic of the USAF resolution target, (b) recorded hologram, (c) reconstructed hologram amplitude, (d) reconstructed phase using approximate phase mask parameters, (e) reconstructed phase using exact phase mask parameters, and (f) residual phase aberration approximation using Zernike polynomials to be subtracted from (e).

shows the recorded hologram. **Figure 6(c)** shows the reconstructed hologram amplitude. **Figure 6(d)** shows the reconstructed phase using approximate phase mask parameters (see the circular fringes). **Figure 6(e)** shows the reconstructed phase using exact phase mask parameters (no circular fringes). **Figure 6(f)** shows the residual phase aberration approximation using Zernike polynomials to be subtracted from (e).

3. Multi-wavelength digital holography (MWDH)

It is well known that one main application of **h**olographic **i**nterferometry (HI) is the generation of a fringe pattern corresponding to contours of constant elevation with respect to a reference plane [2]. These contour fringes can be used to determine the shape of a macroscopic or microscopic three-dimensional object.

There exist three main techniques to create holographic contour interferograms: (a) The two-illumination-point method, (b) the two-refractive-index technique, which is generally not practical because we have to change the refractive index of the medium where the object is located, and (c) Multi-wavelength method, which was adopted in this section [6, 8, 11]. For large height profiles (larger than several microns) 2D topography using single wavelength holographic approach is not appropriate since phase unwrapping has limitations especially for sharp edge variations. As shown in **Figure 7**, the axial displacement of an image recorded with wavelength λ_1 and reconstructed with another wavelength λ_2, with respect to the image recorded and reconstructed with λ_2, is [2, 6, 15–23].

$$\Delta d_z = z\frac{|\lambda_1 - \lambda_2|}{\lambda_2}. \tag{19}$$

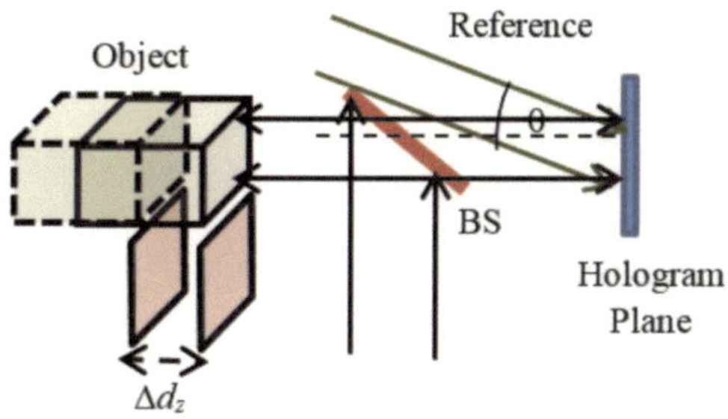

Figure 7.
The path difference of the light rays on their way from the source to the surface and from the surface to the hologram.

This means that the phase shift depends on the distance z between the object and the hologram plane. The height jump between two adjacent fringes in the reconstructed image is

$$\Delta H = z(\Delta\varphi = (n+1)\times 2\pi) - z(\Delta\varphi = n\times 2\pi) = \frac{\lambda_1\lambda_2}{2|\lambda_1-\lambda_2|} = \frac{\Lambda}{2}, \tag{20}$$

where Λ is known as the synthetic wavelength. For larger deformations along the z direction, the phase changes could be hundreds of multiples of 2π. Such large fringe densities may lead to difficulty in determining the object phase using the single wavelength technique. However, multi-wavelength illumination encodes the object height in terms of 2π multiples of the synthetic wavelength, which is generally much longer than either fundamental wavelength. This allows larger object deformations to be measured by multi-wavelength illumination as if illuminated by the single wavelength method, where the "single" wavelength is now given by the synthetic wavelength Λ. Typically, synthetic wavelengths can range from few microns to 10's of microns [16, 18]. The topographic resolution is typically on the order of 1/100 of Λ and the vertical measurement range can reach several Λ's by employing phase unwrapping for heights larger than Λ [24]. However, much longer or shorter synthetic wavelengths to measure millimeter-scale features can also be performed. **Figure 8** shows the advantages of using MWDH to extend the vertical measurement range without phase ambiguity [6]. MWDH may be used to quantify surface topography and displacement measurements for both fixed objects and time-varying objects [17, 18]. It is worth noting that DHM has a lot of applications in living cells [25–30], neural science [31], tissue analysis [32], particle tracking [33–36], and MEMS analysis [37–39].

In multiwavelength DH, both holograms are reconstructed separately at the correct fundamental wavelengths, λ_1 or λ_2. From the resulting reconstructed complex amplitudes $\Gamma_{\lambda_1}(\xi,\eta)$ and $\Gamma_{\lambda_2}(\xi,\eta)$ the phases are calculated as:

$$\varphi_{\lambda_{1,2}}(\xi,\eta) = \arctan\left(\mathrm{Im}\Gamma_{\lambda_{1,2}}(\xi,\eta)/\,\mathrm{Re}\,\Gamma_{\lambda_{1,2}}(\xi,\eta)\right). \tag{21}$$

The synthetic wavelength phase image is now calculated directly by pixel-wise subtraction of the fundamental wavelength hologram phases

$$\Delta\varphi = \begin{cases} \varphi_{\lambda_1} - \varphi_{\lambda_2} & \text{if } \varphi_{\lambda_1} \geq \varphi_{\lambda_2}, \\ \varphi_{\lambda_1} - \varphi_{\lambda_2} + 2\pi & \text{if } \varphi_{\lambda_1} < \varphi_{\lambda_2}. \end{cases} \tag{22}$$

This phase map is equivalent to the phase distribution of a hologram recorded with the *synthetic wavelength*

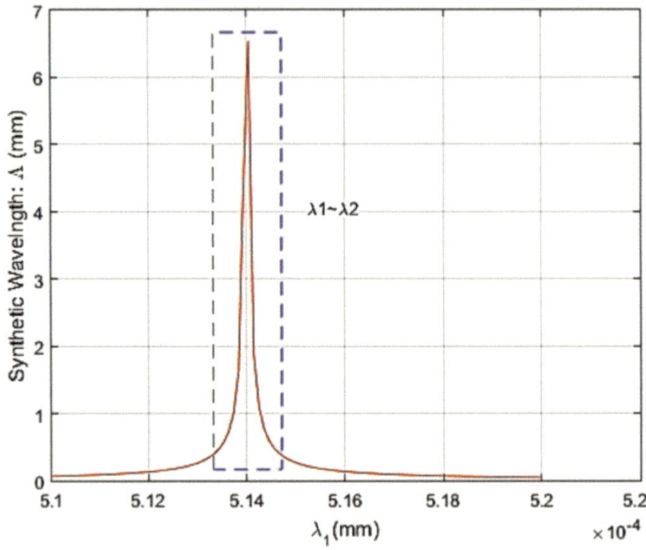

Figure 8.
The advantage of MWDH is that it extends the vertical measurement range without phase ambiguity.

$$\Lambda = \left[\frac{\lambda_1\lambda_2}{|\lambda_1 - \lambda_2|}\right]. \tag{23}$$

At normal incidence, a 2π phase jump corresponds to a height step of $\Lambda/2$, and the change in longitudinal distance or height Δz is given by [2, 6, 9].

$$\Delta z = \left(\frac{\Delta\varphi}{2\pi}\right)\frac{\Lambda}{2} = \frac{\Delta\varphi}{2\pi}\left[\frac{\lambda_1\lambda_2}{2|\lambda_1 - \lambda_2|}\right] = \left(\frac{\Delta\varphi}{2\pi}\right)\Delta H. \tag{24}$$

Note that the transverse resolution is the same as in DH namely, $\Delta\xi = \lambda d/N\Delta x$ is the reconstructed pixel size. Similar to DH, MWDH setup can be constructed using Mach-Zehnder or Michelson configuration as shown in **Figure 9(a)** and **(b)**, respectively. According to **Figure 9(c)** and **(d)**, we notice that the true height measurements in the Mach-Zehnder and Michelson configurations are:

$$\Delta z_{True,Mach-Zender} = \left(\frac{\Delta\varphi}{2\pi}\right)\frac{\Lambda}{2}\cos\theta, \tag{25}$$

$$\Delta z_{True,Michelson} \approx \left(\frac{\Delta\varphi}{2\pi}\right)\frac{\Lambda}{2}, \tag{26}$$

respectively.

One important detail that must be considered when applying the two wavelengths technique is pixel matching. Recall from Eq. (3) that the pixel resolution $\Delta\xi$ of each hologram is dependent upon the fundamental recording wavelength (λ_1 or λ_2). In order for the reconstruction to be successful, the subtraction described by Eq. (18) must be performed on a pixel-by-pixel basis, in which the pixel sizes match between each hologram (i.e. $\Delta\xi_1 = \Delta\xi_2$). This can be accomplished by zero-padding the holograms to alter the numerical resolution according to the following procedure: One hologram is zero-padded prior to reconstruction such that its value of $\Delta\xi$ matches that of the second hologram. The second hologram is then either zero-padded after reconstruction, or the first hologram (which is now larger) is

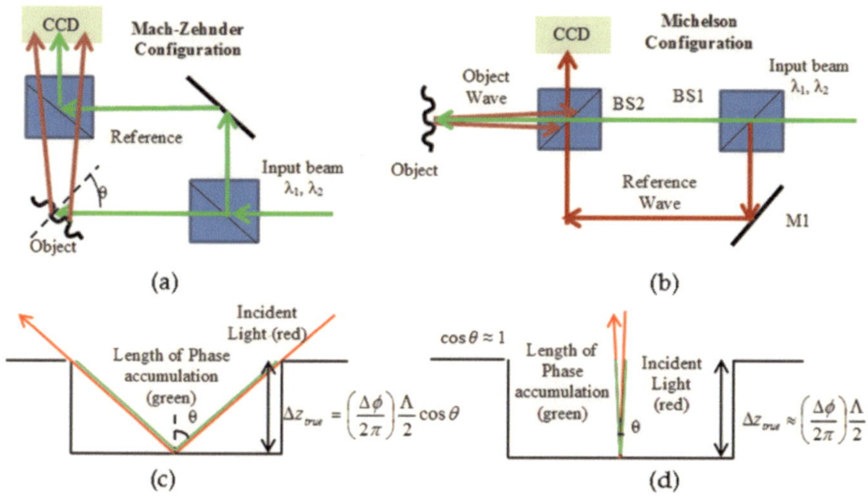

Figure 9.
(a) Mach-Zehnder configuration, (b) Michelson configuration, with illustration of the true height Δz_{True} *relative to the path of phase accumulation for (c) Mach-Zehnder and (d) Michelson configurations.*

cropped, such that the total sizes of each image are again equal. If it is assumed that $\lambda_1 > \lambda_2$ then the degree of padding applied to both the λ_1 hologram pre-reconstruction and the λ_2 hologram post-reconstruction is

$$pad\ size = round\left[\frac{N}{2}\left(\frac{\lambda_1}{\lambda_2} - 1\right)\right], \tag{27}$$

where *pad size* is the number of zero elements to be added symmetrically to each edge of the hologram matrix, rounded to the nearest integer value.

3.1 Experimental results for MWDH with and without spatial heterodyning (MWDH-SH)

An example of a 3D profile setup using the MWDH technique is shown in **Figure 10**. Since the two holograms are recorded sequentially for each wavelength, this technique needs two sequential CCD recordings (i.e. two "shots"). Obviously, this "two-shot" method will not work for dynamic objects. **Figure 11(a)** shows the

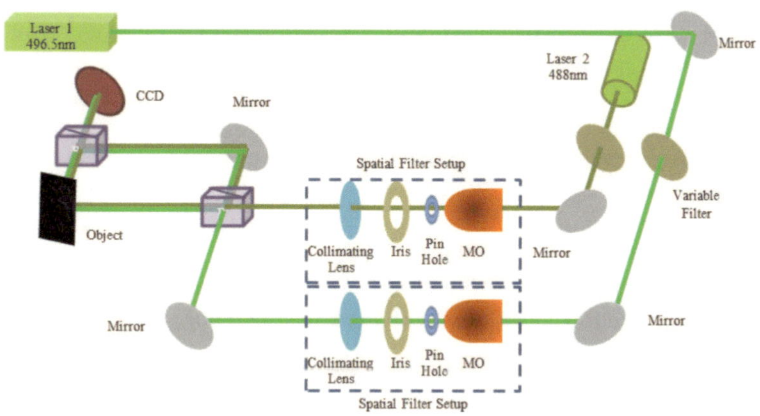

Figure 10.
Shape measurement using MWDH.

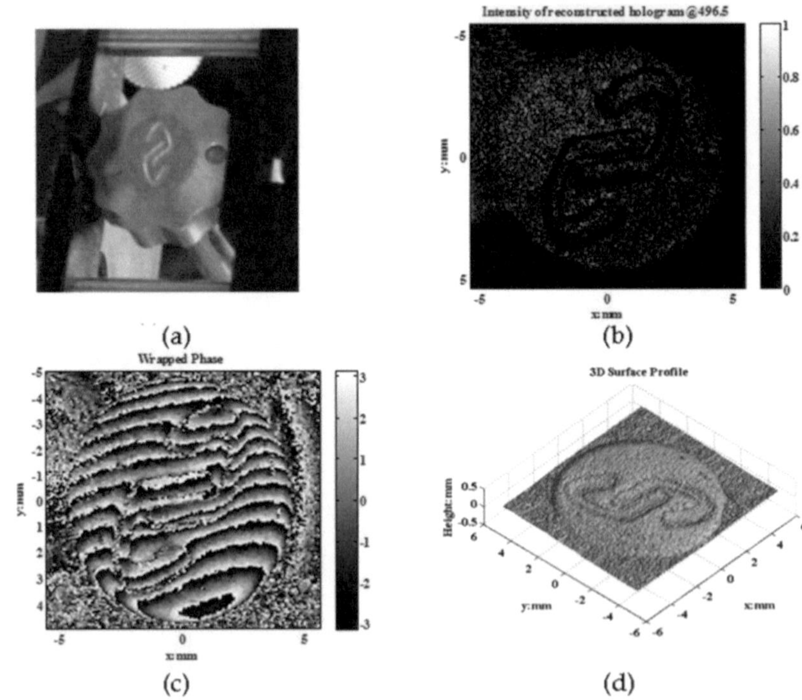

Figure 11.
(a) Newport logo, (b) reconstructed hologram at λ_1 = 496.5 nm, (c) wrapped phase, and (d) unwrapped phase or 3D surface profile. The two wavelengths used are: λ_1 = 496.5 nm and λ_2 = 488 nm and the synthetic wavelength is Λ = 28.5 μm.

Newport logo test object while **Figure 11(b)** shows the reconstructed hologram of the Newport Logo at one of the wavelengths used (λ_1 = 496.5 nm). **Figure 11(c)** shows the wrapped phase and **Figure 11(d)** shows the unwrapped 3D surface profile.

Spatial heterodyning technique has the ability to capture both wavelength measurements in a single composite holographic exposure [6, 9, 21]. This is accomplished by introducing a different angular tilt to the λ_1 and λ_2 reference beams. These angular tilts in the spatial domain introduce linear phase shifts in the frequency domain of the recorded composite hologram. When reconstructed, the different phase shifts result in spatially separated object locations in the image that each correspond to their respective λ_1 and λ_2 recordings. One of these reconstructed images is cropped and digitally overlaid upon the other to perform the required phase subtraction. A typical recording configuration using MWDH-SH method is shown in **Figure 12**. Since the two holograms are recorded for each wavelength at the same time using spatial heterodyning this technique needs only one CCD exposure (i.e. "one-shot") [6, 9, 21, 22]. This method is well suited for dynamic objects which change relatively quickly and only limited by the integration time of the CCD. **Figure 13** shows the reconstructed Newport logo test object. The reconstructed image resolution $\Delta\xi$ = 32 μm/pixel. Note that, two separate reconstructions are required (λ_1 and λ_2) from the single hologram, although only one reconstruction is shown here.

In order to align the two phase images, a block matching algorithm (BMA) is used. After cropping the two reconstructed holograms it is necessary to slide the reference image over the target image looking for best correlation, this is shown in **Figure 14**. Given the typically rapid variation in object phase, BMA algorithms can

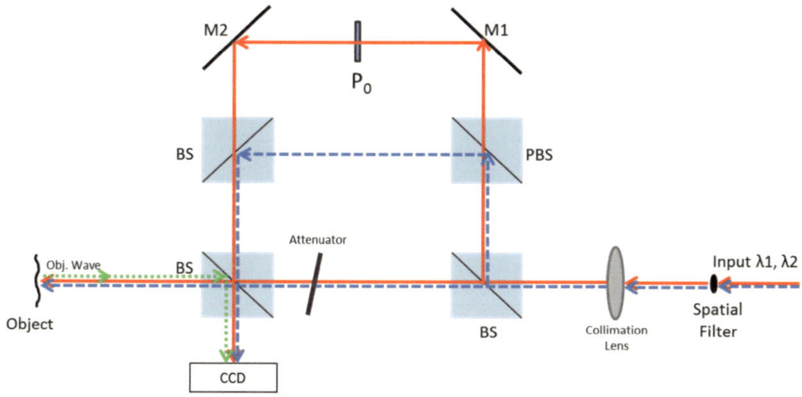

Figure 12.
Lab setup (Michelson configuration) for macroscopic, spatial heterodyne MWDH using coaxial beams and a single spatial filter and collimation lens. The collimation lens should ideally be achromatic at the λ_1 and λ_2 wavelengths. M1, M2: Mirrors, BS: Beam splitter, PBS: Polarizing beam splitter. The polarizer, P_0, ensures λ_1 and λ_2 maintain orthogonal polarization [6, 23].

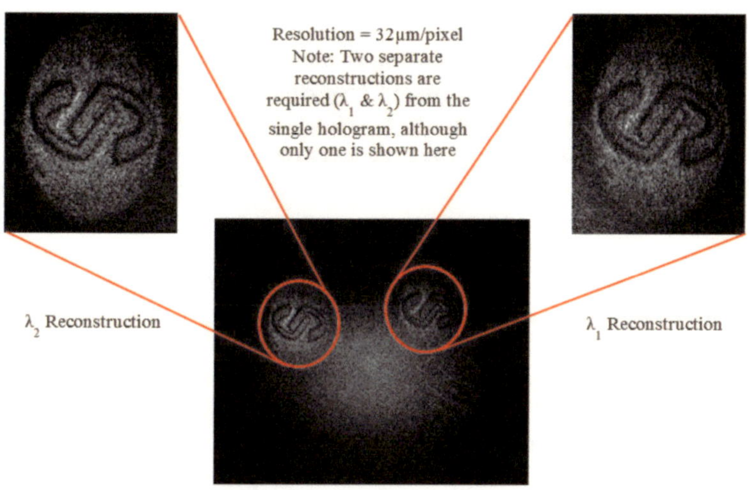

Figure 13.
MWDH-SH hologram reconstruction.

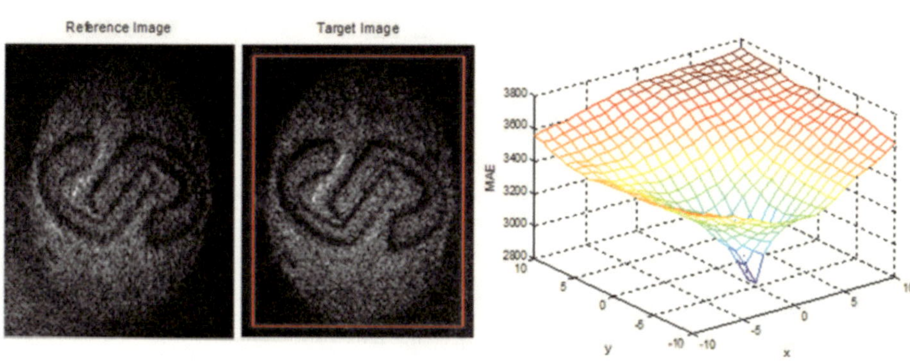

Figure 14.
BMA slides (a) reference image over (b) the target image, (c) correlation. The best correlation occurs at: X_{shift}: 0 and Y_{shift}: −4 pixels.

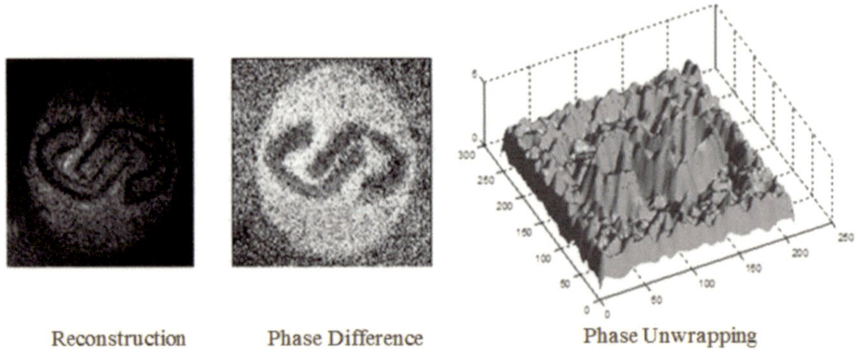

Figure 15.
MWDH-SH phase reconstruction, (a) amplitude reconstruction, (b) phase difference, and (c) phase unwrapping for $\Lambda = 150\ \mu m$.

only match to within ½ pixel. Hence, BMA matching will generally underperform the two-shot method. After aligning the images, the phase difference is calculated by phase subtraction similar to the two shot technique. The example shown in **Figure 15** is for synthetic wavelength $\Lambda = 150\ \mu m$.

An alternative method of matching the two images is to introduce a phase "tilt" to either one, or both holograms during reconstruction which causes lateral shifts in the position of each image. This is typically referred to as introducing a phase mask, $\Psi(m, n)$, during reconstruction, and in general can take any form, although the most commonly used are tilt phases and lens phases. Proper selection of the phase mask (typically found via multiple iterations) can position one hologram reconstruction directly over the other, and phase subtraction may then be performed in a matter analogous to the "two-shot" method previously described, including appropriate resolution matching via zero-padding. An example of phase due to tilt and due to the MO are given by Eq. (13). The hologram matrix is simply multiplied by the phase mask, Ψ, prior to reconstruction. Although the phase mask method is typically more difficult to implement, requiring multiple iterations to arrive at the correct phase mask, it does not suffer from the inherent mismatch error of up to ½ pixel, as the BMA process does. However, the overlap accuracy will now depend upon the accuracy of the modeled phase mask.

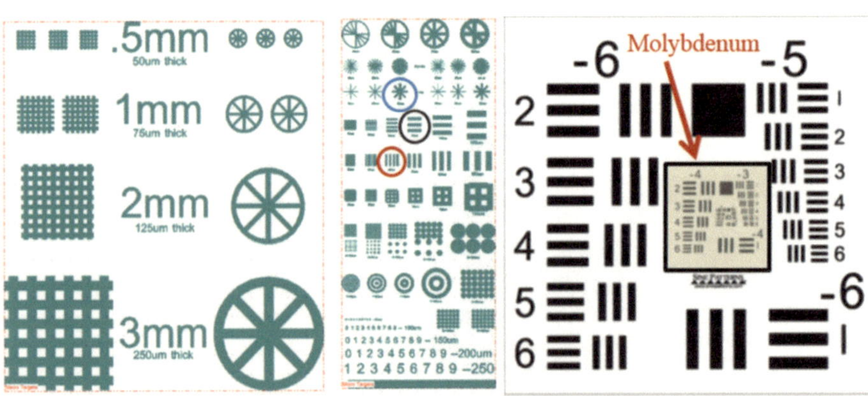

Figure 16.
(a) Custom fabricated micro-scale objects and (b) 1951 USAF resolution chart with a ~ 50 nm reflective molybdenum film sputtered on it.

3.2 Experimental results for multi-wavelength DHM (MWDHM) with and without spatial heterodyning (MWDHM-SH)

In this section, we show an example of using the MWDH technique using a microscopy setup similar to the single wavelength DHM shown in **Figure 4**. Thus, the technique would be abbreviated as (MWDHM). A series of micro-scale objects have been custom fabricated for this experiment as shown in **Figure 16(a)** and **(b)**. **Figure 17(a)** shows a MWDHM setup with achromatic optics and can be operated in either the "one-shot" or "two-shot" Michelson configuration. **Figure 17(b)** is a photograph of the object consisting of 4 bars of photoresist (See element in red

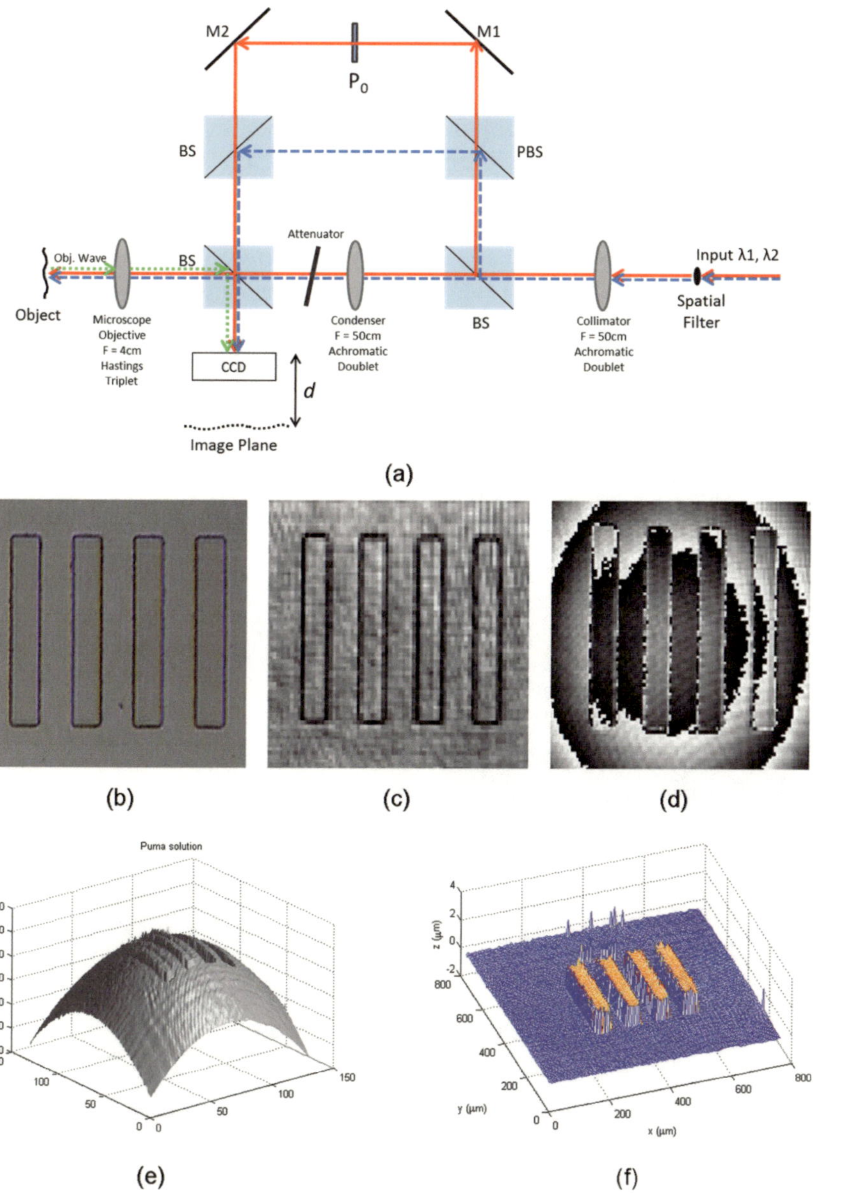

Figure 17.
(a) MWDHM Michelson recording configuration. (b) Photograph of the object. (c) Amplitude reconstruction of the λ_1 hologram only. (d) Wrapped phase difference between λ_1 and λ_2 reconstructions. (e) Unwrapped phase showing MO curvature. (f) the flattened topogram after removal of the curvature.

circle in **Figure 16(a)**), each 50 μm wide, on a silicon wafer substrate. **Figure 17(c)** is the intensity reconstruction of the λ_1 hologram only. **Figure 17(d)** shows the wrapped phase difference between λ_1 and λ_2 reconstructions. **Figure 17(e)** shows the unwrapped phase (Ref. [24]) with the residual MO quadratic phase curvature and **Figure 17(f)** is the 3D topogram after removal of the MO phase using Eq. (13). The relevant reconstruction parameters are: d = 22.7 cm, λ_1 = 632.8 nm, λ_2 = 488.0 nm, Λ = 2.13 μm, Δx = 6.7 μm, and N = 1024, and M = 2.75. Note that the phase rings are due to a slight mismatch in collimation between the λ_1 and λ_2 beams, which causes circularly symmetric phase beating since at least one wavefront is not well collimated. This situation arises often in physical lab setups in which both beams are coaxially aligned and filtered using the same pinhole prior to using a single collimation lens. Chromatic dispersion will prevent both wavelengths from being collimated simultaneously, unless an achromatic lens is used.

Here we show an example using the MWDHM technique with the microscopy setup of **Figure 17(a)**, operated in the spatial heterodyne configuration (MWDHM-SH). The object is a set of 3 rectangular photoresist bars, each 75 μm wide, on a silicon wafer (See element in black circle in **Figure 16(a)**). In this case, the object is simultaneously illuminated by two wavelengths at normal incidence and only a single composite hologram is recorded by the CCD (i.e. "one-shot"). The single hologram is reconstructed twice, one at each fundamental wavelength, and the block-match algorithm is used to align the images prior to phase subtraction. **Figure 18(a)** shows the intensity reconstruction of the λ_1 hologram only, with the region of interest circled, **Figure 18(b)** shows the wrapped phase difference between λ_1 and λ_2 reconstructions after block matching and phase subtraction,

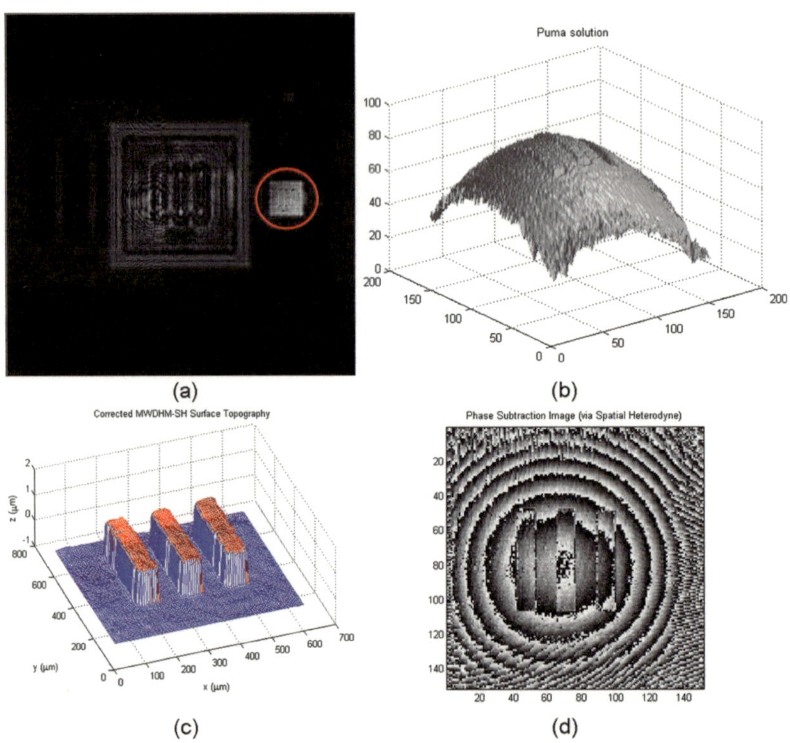

Figure 18.
(a) The intensity reconstruction of the λ_1 hologram only, with the region of interest circled. (b) the wrapped phase difference between λ_1 and λ_2 reconstructions after block matching and phase subtraction. (c) the unwrapped phase with the residual MO quadratic phase curvature, and (d) the 3D topogram after removal of the MO phase, and correction of phase errors.

while **Figure 18(c)** shows the unwrapped phase with the residual MO quadratic phase curvature, and **Figure 18(d)** is the 3D topogram after removal of the MO phase and correction of phase errors. The relevant reconstruction parameters are: d = 23 cm, λ_1 = 632.8 nm, λ_2 = 488 nm, Λ = 2.13 μm, Δx = 6.7 μm, N = 1024, and M = 2.75.

4. Theoretical background of telecentric systems

In a conventional lens systems the magnification changes with object position change, the image has distortion, perspective errors, image resolution loss along the field depth, and edge position uncertainty due to object border lighting geometry. However, a telecentric system, such as the one shown in **Figure 19**, provides nearly constant magnification, virtually eliminates perspective angle error (Object with large depth will not appear tilted), and eliminates radial and tangential distortion. In a bitelecentric system, both the entrance pupil (EP) and exit pupil (XP) are located at infinity. Given that double telecentric systems are afocal, shifting either the image or object does not affect magnification.

As shown in Sections 2 and 3, traditional DHM systems record a digital hologram using a MO. The object phase recovered from digital reconstruction using the Fresnel transform suffers from a parabolic phase factor introduced by the MO. The phase of the MO is superposed over the object phase, often obscuring it. Also, the phase tilt introduced by the reference beam results in linear fringes with high frequency that also obscure the real phase of the object. Numerical techniques as well as optical configurations are usually employed to compensate for both the parabolic phase curvature and the phase tilt. One well-known technique discussed in Section 2 is based on phase mask during reconstruction, which requires knowledge of the setup parameters [13, 14, 40–42]. If the object parameters are unknown a two-step method is used, in which the hologram of a flat reference surface is initially recorded, and upon reconstruction it is subtracted from that of the hologram of the real object [43]. In this section, we adopt two telecentric configurations in reflection and transmission modes to remove optically, instead of numerically, the phase curvature due to MO [44–46]. This telecentric setup can be used in a single wavelength or multiwavelength DHM configurations. It is worth noting that while operating in the nontelecentric mode, *a posteriori* numerical methods will not eliminate the phase aberration completely, as it depends on sample location in the field of view (FOV) [45].

In traditional DHM, the recorded wavefront on the CCD includes the interference of the reference wavefront and the total object wavefront. The total object phase consists of the defocused object phase on the image plane as well as the

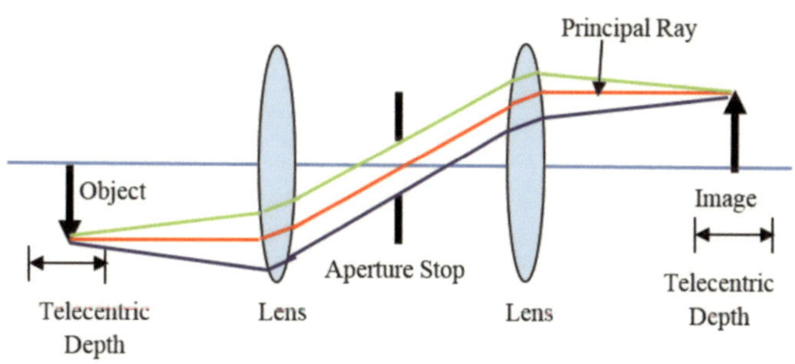

Figure 19.
A double telecentric system.

spherical (quadratic in paraxial approximation) phase due to propagation of the object wave from the image plane to the CCD. The object phase is expressed as [11, 46]:

$$\varphi(x,y) = \frac{jk}{2R}\left(x^2 + y^2\right) + \varphi_{ob}(x,y), \tag{28}$$

where R is the radius of curvature of the spherical curvature.

Typical multiwavelength DHM setups using telecentric configurations in reflection and transmission modes are shown in **Figure 20(a)** and **(b)**, respectively. In each setup, the telecentric system is formed by employing two achromatic lenses and an aperture stop similar to **Figure 19**. The achromatic lenses are crucial to eliminate achromatic aberration due to the use of multi-wavelength illumination. The telecentric system is set in an afocal configuration, where the back focal plane of L_1 coincides with the front focal plane of L_2 $\left(f_1 \equiv f_2\right)$, with the object placed at the front focal plane of L_1, resulting in the cancelation of the spherical phase curvature normally present in traditional DHM systems.

Hence, the 3D amplitude distribution in the image space will be a scaled defocused replica of the 3D amplitude distribution of the object space due to the

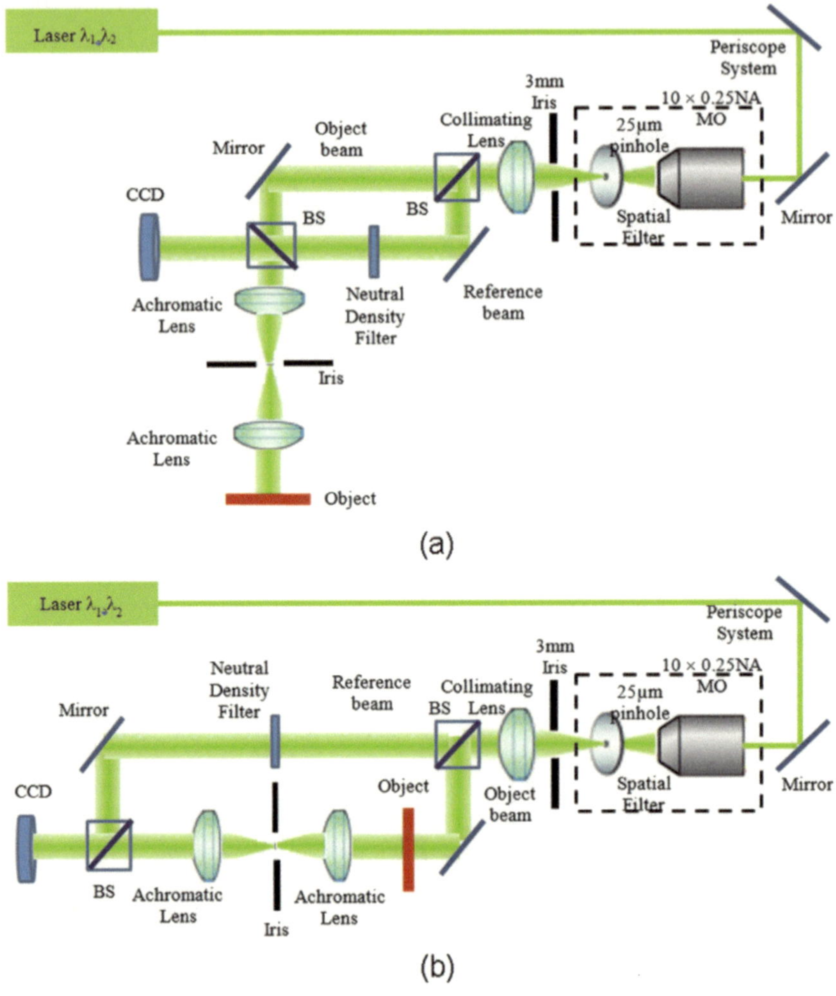

Figure 20.
Schematics of the TMWDHM setups: (a) reflection and (b) transmission configurations.

convolution with the PSF of the lens system. For each wavelength (λ_1, λ_2), the object wave recorded by the CCD can be expressed as [45]:

$$O(x,y) = -\frac{1}{M}\exp\left[jk_{1,2}(2f_2 + f_1)\right] \times \left[O'\left(\frac{x}{M}, \frac{y}{M}\right) * \tilde{P}\left(\frac{x}{\lambda_{1,2}f_2}, \frac{y}{\lambda_{1,2}f_2}\right)\right], \quad (29)$$

where, $O' * \tilde{P}$ is the convolution of the complex amplitude scattered by the object and the PSF, (*) is the convolution operator, and the magnification is $M = -f_2/f_1$ [47].

4.1 Experimental results for single wave telecentric DHM (TDHM)

Figure 21 shows a custom developed user-friendly graphical user interface (GUI) for the single wave reflection Telecentric DHM (TDHM) setup similar to that shown in **Figure 20(a)**. The target object is shown in **Figure 16(b)**.

The MATLAB GUI is connected to a Lumenra LU120M CCD camera using a USB cable. The GUI is equipped with all the parameters needed to adapt to different CCD camera pixel size, laser wavelength, reconstruction distance, reflection vs. transmission mode. In this example, the laser wavelengths used is $\lambda = 488nm$ the CCD pixel size is 5.2 μm, the reconstruction distance is d = 20.2 cm. The reconstructed height is around 120 nm. It's worth noting that slight aberrations due to the optical components exist in the final computed phase. This can be automatically corrected by subtracting the reconstructed phase shape from the background phase using Zernike polynomial approximation of the residual phase as shown in the GUI.

The telecentric technique has a lot of advantages compared to a standard DHM system since the reconstruction parameters in a standard DHM are hard to obtain and need to be measured precisely to obtain the 3D phase information.

4.2 Experimental results for telecentric multi-wavelength DHM (TMWDHM)

Figure 22 shows the GUI for the reflection configuration shown in **Figure 20(a)**. The target in this experiment is a transmissive object (PMMA) on a reflective Si background (See element in blue circle in **Figure 16(a)**). The laser wavelengths used

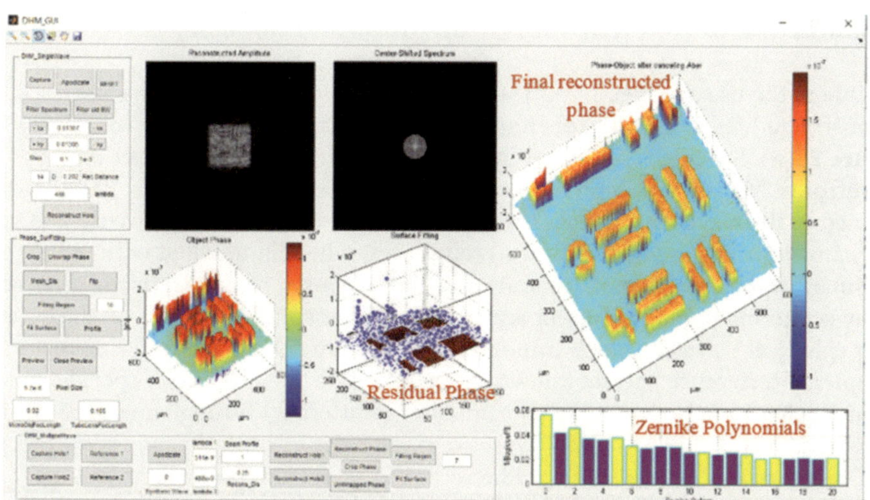

Figure 21.
*A custom-designed GUI showing the TDHM in reflection configuration. The object is a reflective object on a reflective substrate (see **Figure 16(b)**).*

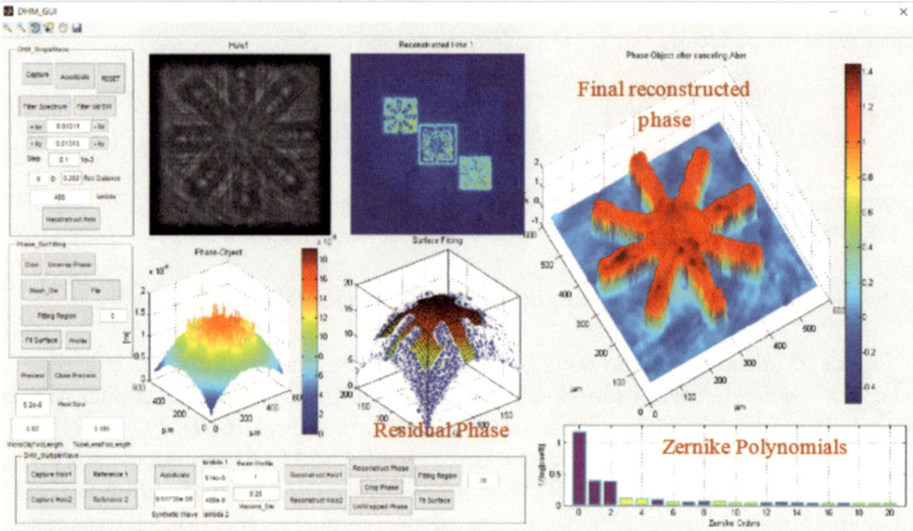

Figure 22.
A custom-designed GUI showing the TMWDHM in reflective configuration. The object is a transmissive object on a reflective substrate (see ***Figure 16(a)****).*

are $\lambda_1 = 514.5nm, \lambda_1 = 488nm$. The synthetic wavelength is: $\Lambda = 9.6\mu m$ and the CCD pixel size is: 5.2 µm. The reconstruction distance is: d = 20.2 cm. It's worth noting that a slight misalignment and/or achromatic aberration may result in one residual fringe to remain in the final computed phase. Although achromatic lenses were used, there still might be some remaining chromatic aberration, since the achromats are not perfect. That might be enough to cause the one remaining fringe of phase curvature. This can be automatically corrected by subtracting a the reonstructed phase shape from the background phase using Zernike polynomial approximation of the residual phase as shown in the GUI.

5. Simulation of coherent speckle on phase and intensity

Due to the use of coherent optical sources, the recorded holograms and reconstructed fields contain coherent speckle patterns, as seen in the inset in **Figure 23(a)**. Speckle is produced by the coherent interference of a set of wavefronts. Mutual interference occurs when coherence is lost, where coherence is defined as the wavefront having constant phase at each frequency. A well-known mechanism for incoherence is optical roughness; when illuminated with monochromatic light the reflected (or scattered) wave consist of the contribution from many scattering points. Different scattering areas or small highlights on the object emit spherical wavelets which combine and interfere coherently resulting in a complex interference pattern known as speckle (Ref. [48–54]). This speckle generation mechanism also applies to transmission (scattering) through an optically rough phase object.

Due to variable phase shifts produced as the wavefront propagates through an optically rough object, the field leaving the object has a corrugated structure of interference. In addition, the presence of an optical diffuser before the object (which consists of small thickness variations) in transmissive configuration, has the same effect as a rough surface in reflective imaging. In this section, we seek to

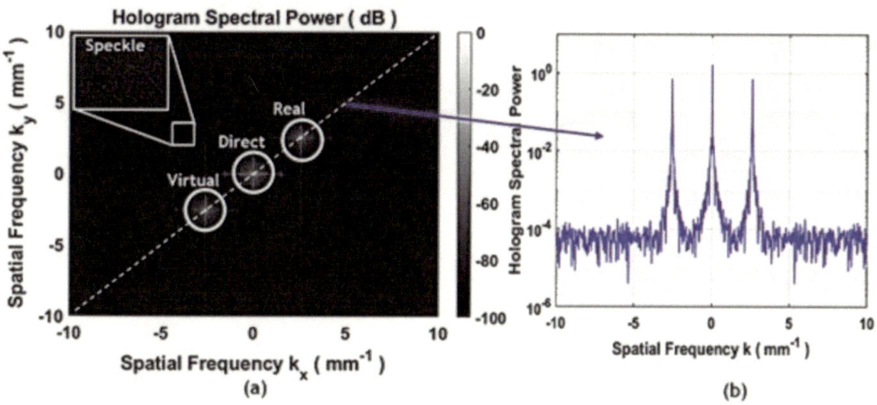

Figure 23.
(a) Spatial frequency spectrum of a hologram recorded using an off-axis digital holographic setup. (b) A slice through constant spatial frequency.

demonstrate an accurate representation of speckle in transmissive imaging through nearly transparent samples, valid for biological imaging applications. To simulate speckle, we consider the complex phasor amplitude $E_O\left(\vec{r}\right) = |E_O|e^{j\left[\left(\vec{k}\cdot\vec{r}\right)+\phi\left(\vec{r}\right)\right]}$ given by a plane wave propagating through an object which induces a spatially dependent phase shift given by $\phi\left(\vec{r}\right)$. A spatially dependent phase shift $\phi_{rough}\left(\vec{r}\right)$ is introduced to the field at the object plane to account for optical roughness/ diffuser. The optical roughness can then be represented by a combination of phasors at each location given by $A_{rough}\left(\vec{r}\right)e^{j\phi_{rough}\left(\vec{r}\right)}$, such that the object complex amplitude wave with the inclusion of optical roughness is given by

$$E_O\left(\vec{r}\right) = E_O\left(\vec{r}\right)A_{rough}\left(\vec{r}\right)e^{j\phi_{rough}\left(\vec{r}\right)} = |E_O|A_{rough}\left(\vec{r}\right)e^{j\left[\left(\vec{k}\cdot\vec{r}\right)+\phi\left(\vec{r}\right)+\phi_{rough}\left(\vec{r}\right)\right]}, \tag{30}$$

where the total phase is the sum of the phase derived from the height profile $\phi\left(\vec{r}\right)$ and the phase introduced by optical roughness. The amplitude contribution of speckle $A_{rough}\left(\vec{r}\right)$ is computed by integrating the absorption coefficient of the material along the optical path length of the roughness. We assume that the phase contribution from each phasor are statistically independent as well as statistically independent from all other phasors such that the phase induced by each surface patch is uniformly distributed over the interval $(-\phi_{max}, \phi_{max})$ (Ref. [49]). The maximum phase shift induced by optical roughness, ϕ_{max}, is derived from the maximum height deviation of the sample roughness. If the surface is rough relative to the optical wavelength, such that each phasor can produce phase shifts of many 2π multiples, the phase shift induced by each surface patch is uniformly distributed over the interval $(-\pi, \pi)$ (Ref. [49]). The numerical propagation of the complex field then captures the coherent interference of the spherical wavelets emitted from the optically rough surface as the wavefront propagates in space.

As an example, we consider a USAF resolution target with a maximum thickness of 10 microns and random height deviations of 1 micron (10% of total height and 1.6λ for red light) due to roughness, imaged through a telecentric holographic configuration with 3x magnification. **Figure 24** shows probability density functions

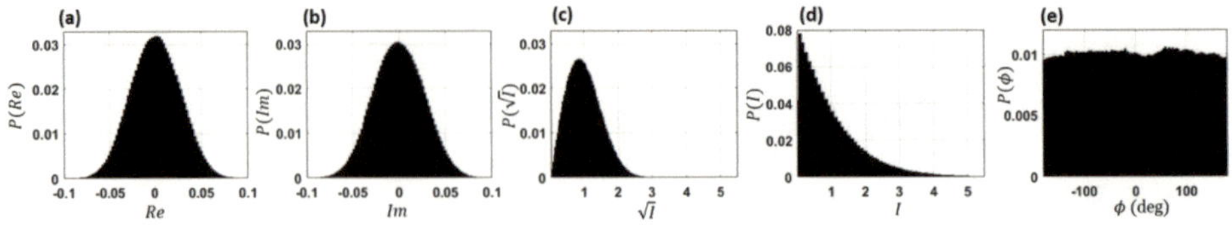

Figure 24.
Probability density functions of reconstructed speckle patterns. The real and imaginary components of the complex speckle field (a,b) are Gaussian distributed, the magnitude and intensity (c,d) are Rayleigh and $\chi_2{}^2$ *distributed respectively, and the phase (e) is uniform.*

computed using the phase reconstruction of simulated speckle patterns; the real and imaginary components of the complex speckle field (A,B) are *i.i.d.* Gaussian random variables, such that magnitude and intensity (C,D) are Rayleigh and $\chi_2{}^2$ (negative exponential) distributed respectively, and the phase (E) is uniform. The validity of the probability density functions in **Figure 24** is well documented in the literature (Ref [48–49]).

In a typical experiment speckle can be reduced using diversity in polarization, space, frequency, or time (Ref. [49]). One of the time domain techniques is through rotating a diffuser or by using a liquid crystal based electronic speckle reducer (Ref. [55]). Another technique is to average multiple holograms or reconstructions recorded by varying the optical path length of the reference beam relative to the object beam (Ref. [56]). **Figure 25** show the reconstructed height profile averaged over increasing phase reconstruction frames, where the initial roughness distributions are assumed to be statistically independent from frame to frame due to the varying optical path length difference between the object and reference beam. **Figure 26** shows the standard deviation of the phase and height profile contribution of simulated speckle as a function of increasing averaging frames. As expected, the standard deviation decreases as $1/\sqrt{N}$ where N is the number of averaged frames.

In this section, we have demonstrated that the distributions of the simulated speckle phase and intensity are consistent with theory and observations in the limit when the optical roughness is large relative to the optical wavelength. In addition, we have shown that the reduction of speckle standard deviation associated with averaging is as expected. While we have demonstrated an accurate and robust numerical representation of optical speckle patterns in holographic imaging, we do not seek to address speckle mitigation techniques in detail. Our goal is to mimic experimentally recorded and reconstructed holograms for realistic machine learning training not to mitigate speckle, as shown in Section 6 below. In future work we seek to explore the sensitivity of speckle statistics to the roughness of the object relative to the optical wavelength.

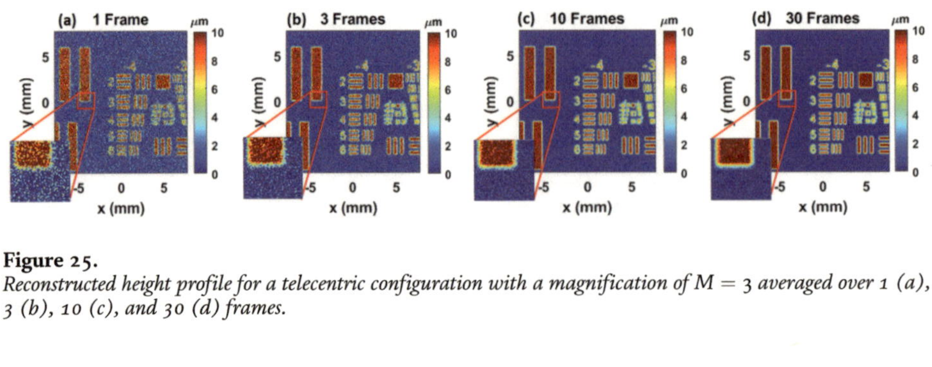

Figure 25.
Reconstructed height profile for a telecentric configuration with a magnification of M = 3 averaged over 1 (a), 3 (b), 10 (c), and 30 (d) frames.

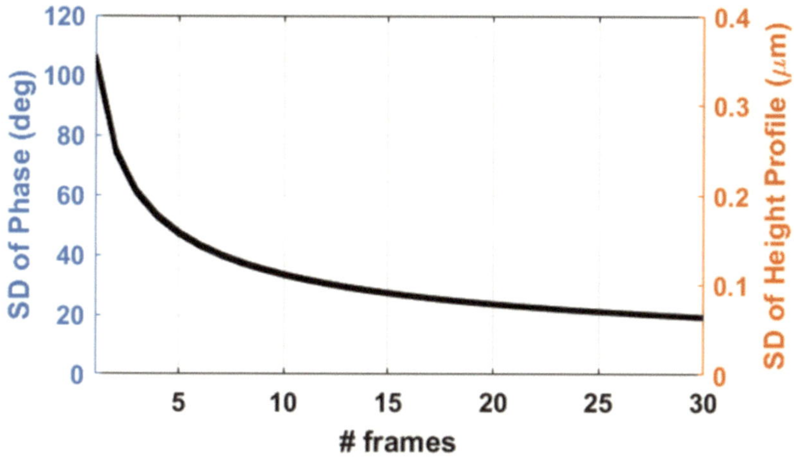

Figure 26.
Standard deviation of speckle phase and corresponding height profile as a function of number of frames averaged.

6. Conclusion

In this Chapter, we developed the theory, the reconstruction algorithms, and discussed the different experimental configurations for digital holography and digital holographic microscopy. We also showed typical experimental setups for single and multiwavelength configurations. We concluded that single wavelength setups are used for heights that do not exceed few microns while multiwavelength-based setups are used for heights that can reach 100's of microns depending on the synthetic wavelength used. We also discussed in details the two shot versus the one shot MWDH setup. Although hologram reconstruction using one-shot setup needs an extra digital correlation step, it is very well suited for dynamic objects which change relatively quickly. We also discussed briefly how Zernike polynomials are used to cancel the residual phase due to the different aberrations in the optical system. We also discussed the theory and experimental setups of novel reflection as well as transmission telecentric digital holographic microscopy configurations. The setup optically removes, without the need of any post-processing, the parabolic phase distortion caused by the microscope objective which is present in a traditional multi-wavelength digital holographic microscope. Without a telecentric setup and even with post-processing a residual phase remains to perturb the measurement. The telecentric technique has a major advantage since the reconstruction parameters needed and hard to obtain in a standard DHM do not need to be measured precisely to obtain the 3D phase information. Finally, a custom developed user-friendly GUI was employed to automate the recording and reconstruction process.

Author details

George Nehmetallah[1*], Logan Williams[2] and Thanh Nguyen[1]

1 Electrical Engineering and Computer Science, Catholic Univ. of America, Washington DC 20064, USA

2 Logan Williams, Air Force Research Laboratory, Wright-Patterson AFB, OH 45433, United states of America

*Address all correspondence to: nehmetallah@cua.edu

References

[1] J. Goodman, *Introduction to Fourier Optics* (Roberts & Company, Englewood 2005).

[2] U. Schnars and W. Jueptner, *Digital Holography: Digital Hologram Recording, Numerical Reconstruction, and Related Techniques* (Springer, Berlin 2010).

[3] U. Schnars and W. Juptner, "Direct recording of holograms by a CCD target and numerical reconstruction," App. Opt., **33**, 179–181 (1994).

[4] L. P. Yaroslavskii and N. S. Merzlyakov, *Methods of Digital Holography* (Consultants Bureau, NY 1980).

[5] U. Schnars and W. Juptner, "Digital recording and numerical reconstruction of holograms," Meas. Sci. Technol., **13**, R85-R101 (2002).

[6] G. Nehmetallah, R. Aylo and L. Williams, *Analog and Digital Holography With MATLAB®*, SPIE Press, Bellingham, Washington, 2015.

[7] T. M. Kreis, M. Adams, and W. P. O. Juptner, "Methods of digital holography: a comparison," Proc. SPIE **3098**, 222–233 (1997).

[8] G. Nehmetallah and P. P. Banerjee, "Digital holographic interferometry and microscopy for 3D object visualization," Frontiers in Opt., **FTuF6** (2011).

[9] G. Nehmetallah, P. P. Banerjee, "Applications of digital and analog holography in 3D imaging," Adv. Opt. and Photon., **4**, 472–553 (2012).

[10] E. Cuche, F. Bevilacqua, and C. Depeursinge, "Digital holography for qualitative phase-contrast image," Opt. Lett., **24**, 291–293 (1999).

[11] E. Cuche, P. Marquet, and C. Depeursinge, "Simultaneous amplitude-contrast and quantitative phase-contrast microscopy by numerical reconstruction of Fresnel off-axis holograms," Appl. Opt., **38,** 6994–7001 (1999).

[12] F. Charrière, J. Kühn, T. Colomb, F. Montfort, E. Cuche, Y. Emery, K. Weible, P. Marquet, and C. Depeursinge, "Characterization of microlenses by digital holographic microscopy," Appl. Opt., **45**, 829–835 (2006).

[13] T. Colomb, J. Kühn, F. Charière, and C. Depeursinge, "Total aberrations compensation in digital holographic microscopy with a reference conjugated hologram," Opt. Exp., **14**, 4300–4306 (2006).

[14] T. Colomb, E. Cuche, F. Charrière, J. Kühn, N. Aspert, F. Montfort, P. Marquet, and C. Depeursinge, "Automatic procedure for aberration compensation in digital holographic microscopy and applications to specimen shape compensation," App. Opt., **45**, 851–863, (2006).

[15] J. C. Wyant, "Testing aspherics using two-wavelength holography," Appl. Opt., **10**, 2113–2118 (1971).

[16] D. Abdelsalam, R. Magnusson, and D. Kim, "Single-shot, dual-wavelength digital holography based on polarizing separation," Appl. Opt, **50**, 3360–3368 (2011).

[17] T. Kreis, *Handbook of holographic Interferometry* (Wiley, Weinheim, 2005).

[18] C. Mann, P. Bingham, V. Paquit, and K Tobin, "Quantitative phase imaging by three-wavelength digital holography," Opt. Exp., **16**, 9753–9764 (2008).

[19] Y. Morimoto, T. Matui, M. Fujigaki, N. Kawagishi, "Subnanometer displacement measurement by averaging of phase difference in

windowed digital holographic interferometry," Opt. Eng., **46**, 025603.1–025603.8 (2007).

[20] P. Hariharan, *Optical Holography; Principles, techniques, and applications*, (Cambridge University Press, Cambridge 1996).

[21] J. Haus, B. Dapore, N. Miller, P. Banerjee, G. Nehmetallah, P. Powers, P. McManamon, "Instantaneously captured images using multiwavelength digital holography," *Interferometry XVI: Techniques and Analysis*, Proc. SPIE 8493, 84930W (2012).

[22] J. Kuhn, T. Colomb, F. Montfort, F. Charriere Y. Emery, E. Cuche, P. Marquet, and C. Depeursinge, "Real-time dual-wavelength digital holographic microscopy with a single hologram acquisition," Opt. Exp., **15**, 7231–7242 (2007).

[23] L. Williams, P. Banerjee, G. Nehmetallah, and S. Praharaj, "Holographic volume displacement calculations via multiwavelength digital holography," Appl. Opt., **53**, 1597–1603 (2014).

[24] J. Bioucas-Dias and G. Valadao, "Phase Unwrapping via Graph Cuts," IEEE Trans., *Image Processing*, **16**, 698–709 (2007).

[25] M. Kim, "Principles and techniques of digital holographic microscopy," *SPIE Reviews* **1**, 018005, (2010).

[26] B. Kemper, D. Carl, J. Schnekenburger, I. Bredebusch, M. Schäfer, W. Domschke, and G. von Bally, "Investigation of living pancreas tumor cells by digital holographic microscopy," J. Biomed. Opt. **11**(3), 034005 (2006).

[27] N. Pavillon, J. Kühn, C. Moratal, P. Jourdain, C. Depeursinge, P.J. Magistretti, and P. Marquet, "Early cell death detection with digital holographic microscopy," PLoS ONE, **7**(1), e30912 (2012).

[28] J. Kühn, E. Shaffer, J. Mena, B. Breton, J. Parent, B. Rappaz, M. Chambon, Y. Emery, P. Magistretti, C. Depeursinge, P. Marquet, and G. Turcatti, "Label-free cytotoxicity screening assay by digital holographic microscopy," Assay Drug Dev. Technol., **11**(2), 101–107 (2013).

[29] J. Kuhn, F. Montfort, T. Colomb, B. Rappaz, C. Moratal, N. Pavillon, P. Marquet, and C. Depeursinge, "Submicrometer tomography of cells by multiple-wavelength digital holographic microscopy in reflection," Opt. Lett. **34** (5), 653–655 (2009).

[30] B. Rappaz, E. Cano, T. Colomb, J. Kühn, C. Depeursinge, V. Simanis, P. J. Magistretti, and P. Marquet, "Noninvasive characterization of the fission yeast cell cycle by monitoring dry mass with digital holographic microscopy," J. Biomed. Opt.,**14**(3), 034049–1-5 (2009).

[31] N. Pavillon, A. Benke, D. Boss, C. Moratal, J. Kühn, P. Jourdain, C. Depeursinge, P. J. Magistretti, and P. Marquet, "Cell morphology and intracellular ionic homeostasis explored with a multimodal approach combining epifluorescence and digital holographic microscopy," J. Biophoto., **3**(7), 432–436 (2010).

[32] K. Jeong, J. J. Turek, and D. D. Nolte, "Fourier-domain digital holographic optical coherence imaging of living tissue," Appl. Opt. **46**(22), 4999–5008 (2007).

[33] N. Warnasooriya, F.Joud, P. Bun, G. Tessier,M. Coppey-Moisan, P. Desbiolles, M. Atlan, M. Abboud, and M. Gross, "Imaging gold particles in living cell environments using heterodyne digital holographic microscopy," Opt. Exp., **18**(4), 3264–3273 (2010).

[34] J. Kühn, E. Shaffer, J. Mena, B. Breton, J. Parent, B. Rappaz, M. Chambon, Y. Emery, P. Magistretti, C. Depeursinge, P. Marquet, and G. Turcatti, "Label-free quantitative cell division monitoring of endothelial cells by digital holographic microscopy," J. Biomed. Opt., **15**(3), 036009 (2010).

[35] H. Sun, B. Song, H. Dong, B. Reid, M. A. Player, J. Watson, and M. Zhao, "Visualization of fast-moving cells in vivo using digital video microscopy," J. Biomed. Opt. **13**(1), 014007 (2008).

[36] C. J. Mann, L. Yu, and M. K. Kim, "Movies of cellular and sub-cellular motion by digital holographic microscopy," Biomed. Eng. Online **5** (21), 1–10 (2006).

[37] Y. Emery, E. Solanas, N. Aspert, A. Michalska, J. Parent, and E. Cuche, "MEMS and MOEMS resonant frequencies analysis by digital holography microscopy (DHM)," Proc. SPIE **8614**, 86140A (2013).

[38] A. Asundi, *Digital Holography for MEMS and Microsystem Metrology*, Wiley, Chichester (2011).

[39] G. Coppola, S. De Nicola, P. Ferraro, A. Finizio, S. Grilli, M. Iodice, C. Magro, and G. Pierattini, "Characterization of MEMS structures by microscopic digital holography," Proc. SPIE **4945**, 71 (2003).

[40] T. Colomb, F. Montfort, J. Kühn, N. Aspert, E. Cuche, A. Marian, F. Charrière, S. Bourquin, P. Marquet, and C. Depeursinge, "Numerical parametric lens for shifting, magnification, and complete compensation in digital microscopy," J. Opt. Soc. Am. A, **23**, 3177–3190 (2006).

[41] F. Montfort, F. Charrière, T. Colomb, E. Cuche, P. Marquet, and C. Depeursinge, "Purely numerical compensation for microscope objective phase curvature in digital holographic microscopy: influence of digital phase mask position," J. Opt. Soc. Am. A, **23**, 2944–2953 (2006).

[42] Z. W. Zhou, Y. Yingjie, and A. Asundi, "Study on aberration suppressing methods in digital micro-holography," Opt. Lasers Eng., **47**, 264–270 (2009).

[43] G. Coppola, G. Di Caprio, M. Gioffré, R. Puglisi, D. Balduzzi, *A. galli*, L. Miccio, M. Paturzo, S. Grilli, A. Finizio, and P. Ferraro, "Digital self-referencing quantitative phase microscopy by wavefront folding in holographic image reconstruction," Opt. Lett., **35**, 3390–3392 (2010).

[44] E. Sánchez-Ortiga, A. Doblas, M. Martínez-Corral, G. Saavedra, and J. Garcia-Sucerquia, "Aberration compensation for objective phase curvature in phase holographic microscopy: comment," Opt. Lett., **39**, 417 (2014).

[45] A. Doblas, E. Sánchez-Ortiga, M. Martínez-Corral, G. Saavedra, and J. Garcia-Sucerquia, "Accurate single-shot quantitative phase imaging of biological specimens with telecentric digital holographic microscopy," J. of Biom. Opt., **19**(4), 046022 (2014).

[46] E. Sánchez-Ortiga, P. Ferraro, M. Martínez-Corral, G. Saavedra, and Ana Doblas, "Digital holographic microscopy with pure-optical spherical phase compensation," J. Opt. Soc. Am. A, **28**, 1410–1417 (2011).

[47] G. Nehmetallah, "Multi-wavelength digital holographic microscopy using a telecentric reflection configuration," Topical Meeting in Digital Holography and Three-Dimensional Imaging (DH), **DM3A.7**, Shanghai, China, 24–28 May (2015).

[48] J. W. Goodman, *Statistical Optics* (John Wiley & Sons, New York, NY, 1985).

[49] J. W. Goodman, "Some fundamental properties of speckle," J. Opt. Soc. Am., 66(11), 1145–1150 (1976).

[50] J. W. Goodman, *Speckle Phenomena in Optics: Theory and Applications* (Roberts & Co, 2006).

[51] H. Funamizu, J. Uozumi, and Y. Aizu, "Enhancement of spatial resolution in digital holographic microscopy using the spatial correlation properties of speckle patterns," Continuum 2 (6) 1822–1837 June(2019).

[52] Y. Park, W. Choi, Z. Yaqoob, R. Dasari, K. Badizadegan, and M. S. Feld, "Speckle-field digital holographic microscopy," Opt. Exp., 17(15), 12285–12292 (2009).

[53] J. Zheng, G. Pedrini, P. Gao, B. Yao, and W. Osten, "Autofocusing and resolution enhancement in digital holographic microscopy by using speckle-illumination," J. Opt. 17(8), 085301-7p.p. (2015).

[54] T. Baumbach, E. Kolenovic, V. Kebbel, and W. Jüptner, "Improvement of accuracy in digital holography by use of multiple holograms," Appl. Opt. 45 (24), 6077–6085 (2006).

[55] J. Boonruangkan, H. Farrokhi, and Y. J. Kim, "Rotational diffuser for speckle reduction in quantitative phase imaging," *2017 Conference on Lasers and Electro-Optics Pacific Rim (CLEO-PR), Singapore, 2017, pp. 1–2, doi: 10.1109/CLEOPR.2017.8118645.*

[56] H. Lin and P. Yu, "Speckle mechanism in holographic optical imaging," *Opt. Exp.*, vol. 15(25), 16322–16327 (2007).

Chapter 5

Application of Augmented Reality in TVET, a Modern Teaching-Learning Technology

Balakrishnan Chandrasekar

Abstract

Technical Vocational Education Training (TVET) in India is increasingly receiving significance with focus towards skill development and employability in the development sector. The skill development and employability of skilled resources depend on the extent of design, delivery, attainment, relevance of vocational training programmes. The objectives of TVET institutions have varied with focus on the type of technical vocational and skill training, the delivery systems, employment potential, and employment in the relevant sectors. They are predominantly relevant to the geographic location, availability potential resources *viz.* manpower, funding and employment scenario. The skill training providers largely face several constraints and difficulties in the delivery systems, teaching-learning process, which have varied with learner's expectations and the education needs. The technology-enabled teaching-learning has directed towards increased use of information communication technology (ICT) and computer-based teaching technologies in the learning centres. The computer-based teaching, virtual experiments and practical-based learning find wider application. Very little has been done in terms of relevance of vocational training and applications. This chapter attempts to review the available augmented reality or extended augmented reality for technical vocational education training and skill development, its relevance and increasing the impact of student leaning application in the skill centres. The author finally shares the experience of use of extended augmented reality for a vocational training.

Keywords: TVET, virtual learning, e-learning, augmented learning technology

1. Introduction

Technical Vocational Education and Training (TVET) is a comprehensive system provided for training at the technician level to aspiring youth, which essentially forms part of an integral employment system in any developing nation. TVET and skill development systems are pivotal to employment creation, employment generation and employability across a variety of domains of economic activities in a country. India has taken a stride in providing wider opportunity for skill training and the technical vocational education for few decades. TVET and skill development have grown manifold with opportunities and seamless integration with formal system of training and employment. Across the Asian region, the technical vocational education has broadly remained the same within the context of skill-based training system, outcomes and learner attainment.

IntechOpen

All developing nations have instilled strong belief that well-trained workforce is the cornerstone to the growing economy and development of the nation. They always serve as a backbone to the nation in bringing more skilled workforce, aspiring to quality services and relevance to the market-driven economy.

In order to give an understanding of evolution of TVET in a typical developing nation like India, a sequential history of policies and strategies are briefly presented in the following section. It is also common to observe that abutting nations to India have systems even before India's republic, as they all have emerged from the erstwhile British regime, countries have adopted as per the country's need, demand and availability of resources with time.

The vocational training system in India can be traced back to the British Rule, the pre-independence and post-independence (before and after 1947, respectively). During the onset of economic growth, the vocational education received little attention. However, the government at that point of time consciously put in place policy initiatives recommended by such Committees during pre-independence era. The key initiatives in technical vocational education training system enumerated schemes detailing the programme framework, learner attainment, etc. The policy decisions in pre-independence were through the statutory committees and intended reports, as follows: (a) Wood Despatch [1], (b) Indian Education Commission [2], (c) Rise of National Education Movement (1905–1921) [3], (d) the Hartog Committee (1929) [4], (e) the Sapru Committee [3], (f) Abbot Wood Report [4], (g) the Zakir Hussain Committee [5] and (h) the Sargent Report and Central Advisory Board on Education [6].

The recommendations and strategies put forth by these committees laid the foundation for design, preparation and institutionalizing the education system inclusive of technical vocational education. The education policy broadly covered the technical education (the engineering stream), technician and vocational system and related areas. The technical vocational system received its significance as per the priorities at that point and funding availability. The section below presents the highlights relevant to development of TVET systems, in the context of the learner engagement and pedagogy.

The formal structure of education in general composed of primary, secondary schooling and pre-university (5 + 3 + 3 years, respectively), and the vocational studies commenced after 11 years of schooling, as recommended by the Sapru Committee [3]. During this phase, the basic education syllabus along with the crafts curriculum was proposed, wherein all school students to receive the vocational curriculum to be imparted at school level. The pedagogy in basic education and vocational education was given importance and began as a policy as also outlined in its report [4]. In this phase, the assessment, evaluation and student retention in classroom for instructional-based learning were also emphasized to be stakeholder needs. Classroom teaching with common instructional medium and the regional language (mother tongue) were given stressed during this period. Following this, the learning resources such as textbooks, lab manuals and workshop resources received attention in the vocational education at the secondary level.

The vocationalization of secondary education was given importance wherein pre-final and final-year school students at the end of 11 years schooling had the option to specialize in technical and vocational training [5]. During this phase, the vocational education faced impediments during implementation including lack funds, adequate training resources in the centres, schools, lack of trained teachers.

A significant change in the education system along with the technical vocational education training was experienced during the post-independence era. The technical and vocational education received its importance due the labour market needs and demand in the industrial manpower. Following this demand, the education received much attention. Several institutional structures along with formal systems in terms of committees and commissions were formed. Some of the key institutional mechanism were (a) University Education Commission [7], (b) Secondary Education Commission [8], (c) Kothari Commission [9], (d) Central Advisory Board of Education [9], (e) National Policy on Education [10], (f) Report of the Working Group on vocationalization [3], (g) Adiseshiah Report – Vocationalization of Higher Secondary Education & 10 + 2 Committee [11], (h) Working Group on Vocationalization of Education [3], (i) National Policy on Education and the Action Programme on Vocational Education (1986) and (j) National Policy on Education 1986, Revised Policy (1992) [12].

It may be seen the aforesaid institutional mechanisms are key to development of TVET system as it is presently and keeps evolving as per the needs, stakeholder demands and the domain expertise. Taking into account the experience of pre-independence, the Kothari Commission [13] consolidated an education system to meet the national needs, addressing the labour market scenario, taking into account best expertise of the global situation. Thus, the emphasis for vocationalization was introduced at the pre-university, which included wide variety of areas including agriculture, engineering, non-engineering trades. These areas helped learners to acquire specialization in these areas at the higher education level. Vocational courses after 10 years schooling was emphasized for rapid expansion. Trade specialities in health sector, administration, commerce and small-scale industries were introduced. The trade certification levels such as certificate courses, diploma and post-diploma were designed, and the same was offered by the institutes owned by central government and the states.

With the growing needs and demands in the formal vocational system, there was need to undertake surveys and reformulate strategies to meet the skill demands at various level, the central government constituted with the aim to uniformly expand the scope of such technical institutions across the states. The expansion of trade certification programmes in the ITIs to meet the industry needs at rural blocks for increasing employment to such youth was recommended in Subanayagam Report, 1978 [3]. It is at this stage, ITIs were given importance and positioned as units of training service providers with the industry standards of curriculum, syllabus, examinations and certification prescribed by a national body for the identified trades. The process of streamlining a statutory body named, National Council for Vocational Education at the central level and concurrently, states also the established State Councils for Vocational Education to meet the state and regional needs. These institutional mechanisms served to set norms, standards for various trade level courses, certification standards and examination systems.

With increasing demand for trained manpower in the labour market, the technical vocational education and training gained more relevance in the national and regional level context. The central government at that point realized the importance of restricting TVET system. The vocational and skill trades in the context of industry demands were addressed in a comprehensive manner. During this phase, few additional institutional mechanisms were recommended by the Kulandaiswamy Report, 1985 [3]. By this stage, the ministry responsible for education was vested with the responsibility to synthesize curriculum, learning resources, assessment system and introduction of vocational education in schools. Thus, the responsible ministry set up a national apex institution, the National Council for Education Research & Training, which grew to an extent to spin-off a dedicated vocational education training and research in vocational curriculum and pedagogy, and develop the trainers programme at the national level, which is now named as Pandit

Sunderlal Sharma Institute for Vocational Education (PSSIVE). This was given complete powers to prepare curriculum in additional to the schooling at all levels for the vocational education and preparing the learning resources.

It is seen that the vocational education and training remain the core for skilling the learners and aspiring youth seeking jobs in the market. The central government have initiated various strategies at all levels to make relevant and appropriate with changing trends, demands and needs in the view of variety of stakeholders. The government at all times has accorded priority to skill development as the central sector scheme to promote TVET in partnership with industry stakeholders [14].

2. Recent developments in technical vocational education and training and skill training system

In the context of this chapter, TVET system relates to a formal training system where students after obtaining school certificate enter the vocational stream leading to a certificate, diploma and specialized trade-related courses. In general, the trained youth entered this formal system and were allowed to do an apprenticeship training attached with industry leading to award of certificate for which the learner is enrolled. The skill development system is a structured system of skill training for youth who have never attended a school, students attended school and later drop-out for reasons, or youth who never had a formal education in any of system and those aspiring to enter the labour market with a certification (recognized by the designated skill certification authorities following such norms), and this certification empower the youth in getting an employment in labour market.

Over the decade, the TVET system emphasized the requirement for change in the organizational structure at various levels that will help to address the core aspects of vocationalization. The efforts for introduction of vocational education at various levels achieved impact to an extent at that point of time. The government at national level has taken series of efforts on its own and with the support of externally aided institutions such as GIZ India, ILO, The World Bank and few private sector partners including NGOs to conduct a review of the TVET system [15].

Some of the key issues in the TVET system are briefly illustrated, which are in general relevant to the developing nations like India.

i. The TVET system shall be dynamic that is responsive and address the needs of the labour market with the growing economic development.

ii. The strategies for improving the relevance and effectiveness of public vocational training institutions shall be the key factor, keeping in view the industry engagement and representative of industry stakeholder across all the policies of a country and keep pace with the demands and requirements.

iii. A well-designed industry-institute engagement model, learner engagement model with the training service provider to meet all times demands will be more appropriate.

iv. The management structure of TVET at all levels needs to be proactive and responsive, which meant defining the roles and functions and be able to address the intended objectives at the national, regional and state levels and other key stakeholders directly responsible.

v. Autonomy to skill training providers in line with the national strategies, industry demands and needs to address the skill requirements and flexibility to engage with such stakeholders to make vocational skills more relevance. The skill training providers to have a localized management structure and representation of industry stakeholders to address skills needs, demands, work in achieving the efficiency of skill training providers, be able to address local and regional needs at time, more towards addressing skill gaps by equipping skill training providers with modern labs that shall address the industry demand skills.

vi. In order to achieve the effective utilization of infrastructure created, national and regional levels, continual improvement of the system is a must; therefore, increase allocation and availability of financing are to be under top priority and should form part of the TVET strategy.

vii. In the present context, synergizing the formal TVET system and skill development system is a discussion point. Both the systems address skills as the learning outcome. Then, the learner engagement model is concern, which varies with the type of trades, institutions where the learner acquires the skills, the context where the learner acquires such skills, for example in the skill centres, industry floor shop, etc.

The vocationalization in India turned to large-scale training system and the need for private sector engagement came into a realization based on the review and assessment by national and international institutions. Studies were conducted by the World Bank, GIZ India, UNESCO-ILO and few joint studies to elicit nature of demand and potential. Consolidating the recommendations and outcome of the above strategic studies, the Central Advisory Board on Education in 2008 [16] stressed on quality, quantity, access and relevance. The objective was also to engage with private sector in large-scale design, development and deployment of skill trades which lead to employment, education and empowerment. The private sector participation and learner engagement model for the large-scale system varied with the type of service provider, the institutional management structure, financial capacity and capability and reaching the needy across demand sectors, needy sectors and employment potential. A system that allowed greater cost sharing, that is moving from a system which is increasingly financed by the private sector and by student fees and ensuring that vocational technical education, has vertical mobility to higher education for the prospective students.

The technical vocational education and training system in India has evolved to meet the labour market needs at various levels. The integration of skills in the formal education system is presented as a comparison as given in **Figure 1**.

The technical vocational education training in many Asian countries such as Thailand is termed as experimental-based learning process [18] and authors also define as 'an education designed and developed to enhance learners' technical skills, human talents, cognitive accepting attitudes and work behaviour in order to make learners employable in industries' [19].

The key characteristics in the TVET system have varied with type of vocational technical training, regional needs, skill demands needed in the employment sector and the extent of handling with type of equipment, devices and the like. The common characteristics include as follows:

a. Contextualized learning: vocational learning content; pedagogy—teaching-learning environment, learning ecosystem; type of workplace; type and level of learners, who make the entire context relevant and appropriate

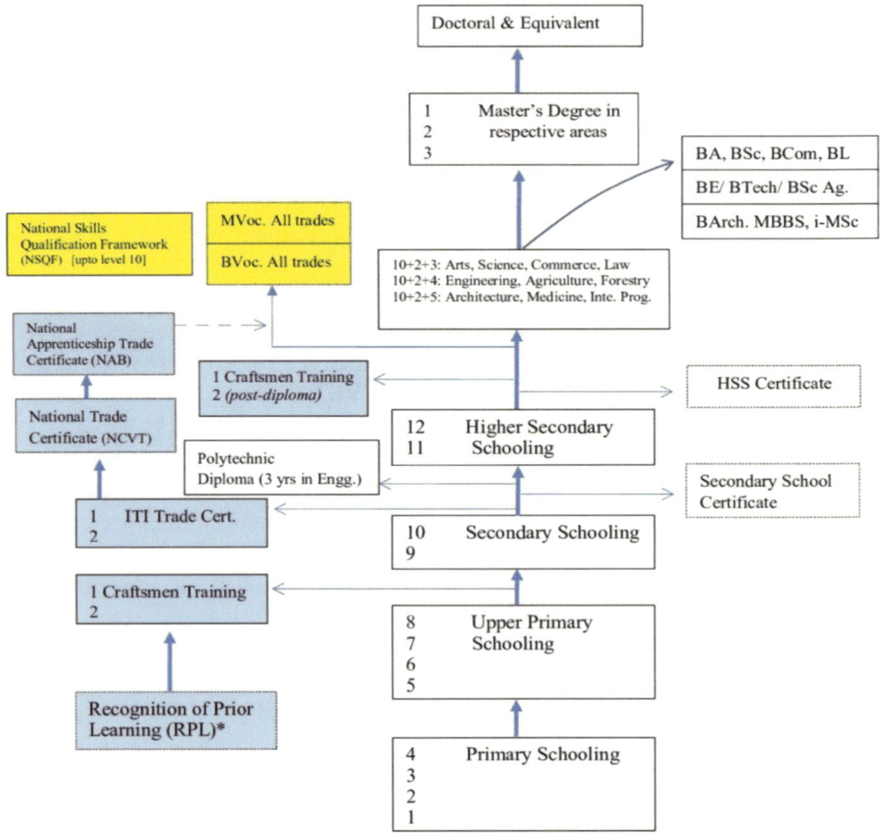

Figure 1.
*TVET & skills training system in India. Source: B. Chandrasekar et al. [12]. *Recognition of prior learning (RPL) is a certification process, which allows learners fortransition from non-formal to an organized market by acquiring skill certification. It is a process of recognizing previous learning, often experimental towards a given qualification and allows horizontal and vertical mobility in the new skill qualification framework [17].*

b. Varying learner characteristics, skill needs and demands in the type of skill programmes

c. Trainers adopting different learning methods, classroom-based teaching, mocks & trials in the classroom/workshops, mock labs and virtual experiments, case study resources, visit & experience on practical handling techniques, etc.

d. On the job training, wherein fresh graduates learning in varied systems and approached lands in shop floor, workshop to get hands-on training and experience to mature into assembly productions, manufacturing, etc.

The structure of TVET common across all Asian nations is illustrated in **Figure 2**.

In general, the Asian nations have similar systems wherein youth entering the labour market have less skills or never gone through a formal training for an identified trade. Thus, the need for a system integrating the learning phase in the vocational training centres and that of the industry need to meet the demands for specific tasks and job roles. This holds true across all developing nations. There is always a wide gap in the demand supply system matching the skills across sectors.

While the characteristics are focussed and clear with type of learning objectives for technical vocational education remains pivotal to employability, the significant role of the system is to provide orientation and initial training in the key skills and upgraded skills/upskilling in the relevant areas of skill and vocational training. A combination of the above two key objectives will lead graduates employable in the stream of training obtained. For the initial technical and vocational training, the beneficiaries (student population, youth) at the age group of 16–18 years have obtained the required schooling education and seek employment in any technical vocational sector. They are employed in the production, manufacturing, shop floor jobs in such units with certain skills levels competent to perform such tasks, those who enter the labour market fresh. Another category of skill and vocational learners includes those who desire to upgrade skills in a variety of areas with varied experiences in their prior learning experience (PLE) or in related fields to upgrade their knowledge and skills and facilitate/help move up in their career. The learning experience- and practical-based learning is analogous to higher education as they are counterpart in their field of learning ecosystem, and the comparison is given in **Table 1**.

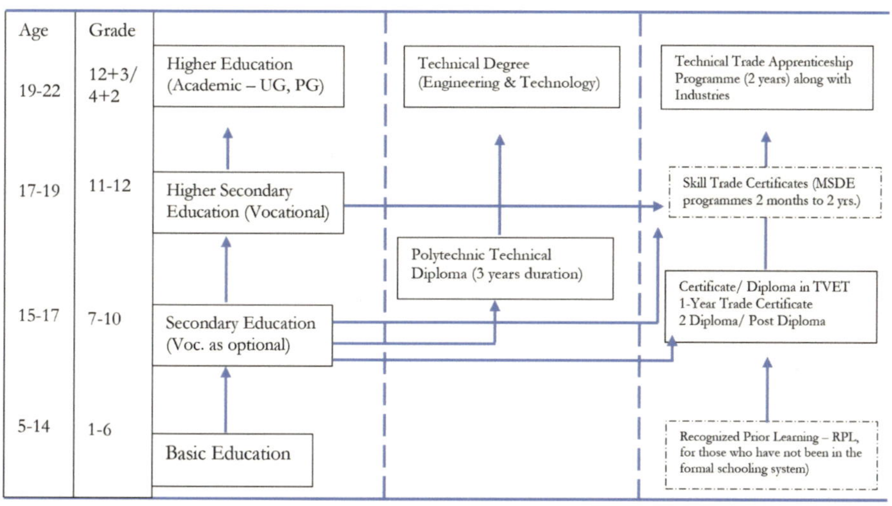

Figure 2.
TVET system in Asia (source: adapted from [20]).

Key attributes in learning systems/processes	TVET	Higher education
Learning environment	Practice-based learning	Fundamental, theory & concept based learning (basic & advanced level)
Learning goals	Occupational competence: competency-based learning, skill-oriented learning	Discipline-based learning, autonomy/ flexibility of choice to specialize in particular areas, upgrade to higher level in education stream; build over all personality
Learning orientation	Demand-based training, labour market needs for such qualified workers	Cannon of representative knowledge from broad based to focussed; research as focus, learn from objectives to evidence based
Knowledge/learning & assessment	Adoption of competency-based education	On the job training and assessment of student knowledge in a grading approach

Table 1.
Key attributes in learning systems and processes a comparison.

3. Vocational learning systems

The learning systems in TVET or skills have largely remained conventional wherein typically, the classroom methods with the concept of technical vocational content imparted to learners, while that of practical handling is also within the classroom using demo equipment and devices. The disciplines largely include engineering, manufacturing and production sector trades.

The duration of classroom learning, use of pedagogy and practical training, practice handling of such devices and equipment varied with type of trades, level of certification and the kind of programmes.

With the increasing demand for employment and employability in a variety of areas covering industry, agriculture, service sector and related areas, the learning systems emerged into a combination of traditional method combined with computer-based or technology-enabled learning. The learning contents using the computer-based have varied with time, type of trade and the extent to which the contents can be delivered using technology platform.

ICT in vocational education became a new concept of learning in early 2000 and significantly taken up various nations to meet the increasing skill needs and demands, while attempting to maintain the quality of classroom delivery equivalent to the conventional face-to-face teaching experience. However, during the initial phase, it is kept to the extent as per the needs and demands in the respective areas. The attainment of skills levels varied with the speciality of trades, level of skill programme and availability and usability of the contents. UNESCO-UNIVOC [20] reported that ICT in vocational technical education had the potential to create transformative changes in skills delivery and attainment to the expected level of employers. The possibilities include flexible, life-long learning, enhanced knowledge sharing and social learning between peers and experts, and this augmented system of virtual learning and simulated experiments will meet the sustainable objectives of TVET learning attainment.

With the increasing availability of computer- and technology-enabled learning, the vocational and skill learning content had focus on delivery of concept, modular-based approaches to skill trades, learners' ability to pick up the needs relevant to the employers' expectations. Taking into account the real-time situation, virtual reality and augmented reality are the latest option, where learners at their desktop can access the real-time learning on virtual mode. This shall have combined with gaming techniques and assessment modules to review the learner skill attainment.

The benefits of augmented reality and virtual reality were reported by Diegmann, Schmidt-Kraepelin, Eynden and Basten in [21], mentioned about improving the safety of experiential learning and teaching, access to contents and resources for complex topics, enhancing learners' interest and keep up the enthusiasm for long-term learning. Later authors mention that virtual reality learning increases learner's motivation and have experience of seeing issues in a real-time situation, at the same time, learners are able to focus on the key aspects to practical handling. The virtual reality or the augmented reality provides support for continual learner engagement and learners are able to mix and match the collaborative learning with more learning abilities.

The pedagogy in the technical vocational education training over the decades can be traced since its gaining significance, which is illustrated in **Figure 3**.

The learning attainment varied with type of combination of technology and disciplines of vocational training by the learners.

A review of unemployment rate in the developing countries indicates that effective learning technologies will address the quality of TVET and employability, effectively the technology-enabled learning can meet largely to the extent and the variety of stakeholders addresses the factor in a variety of means. From the data, it is seen that a large potential of female population are still untapped and potential for vocational training in new areas is still to be explored. Efforts are taken for enhanced participation of female population across the Asian nations.

Unemployment rate of youth total and female in the Asian region, as estimated by the World Bank 2019 is as under as follows:

		Bangladesh	Sri Lanka	India
Total	2015	10.79	19.26	22.33
	2019	11.87	21.23	23.33
Female	2015	13.16	25.80	23.14
	2019	15.42	29.82	24.31

Source: [22].

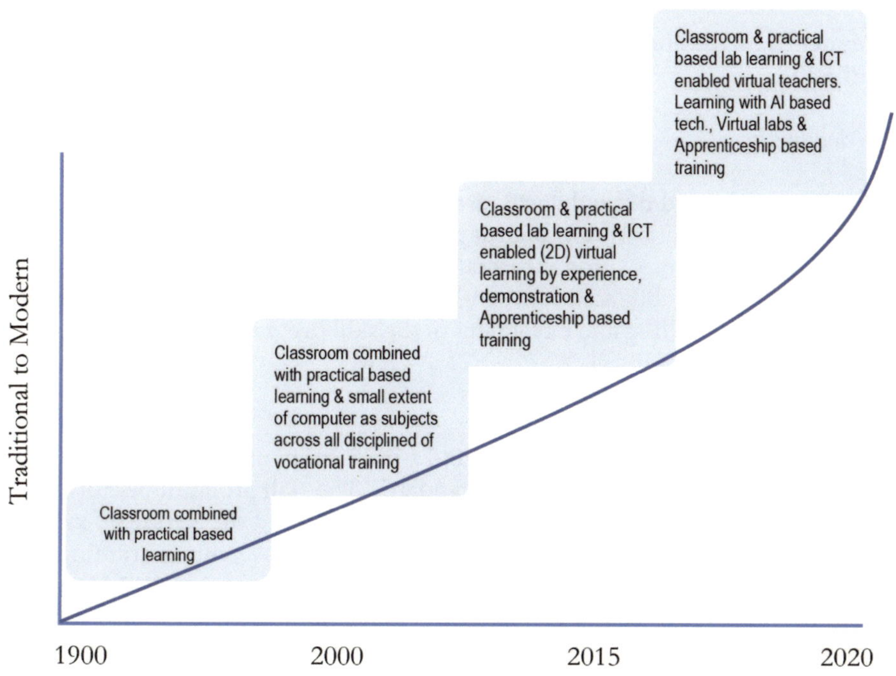

Figure 3.
Evolution of teaching technologies in TVET.

4. Learner engagement model

As mentioned, the learner engagement model varies with the type of vocational training institutes either operated and management by the responsible authorities. In a research study conducted by the author for skills development in private sector initiatives, the vocational training institute (or) the vocational service provider is engaged from the end-to-end services in the skill development processes. The service provider is responsible for mobilizing the learners, who are selected based on a prior skill obtained, earlier learning acquired or on the knowledge and skills in the specific domain areas. These learners undergo minimum learning hours, which are usually combination of real classroom teaching and virtual training using multimedia learning resources. These are generally 2D or 3D learning resources. These

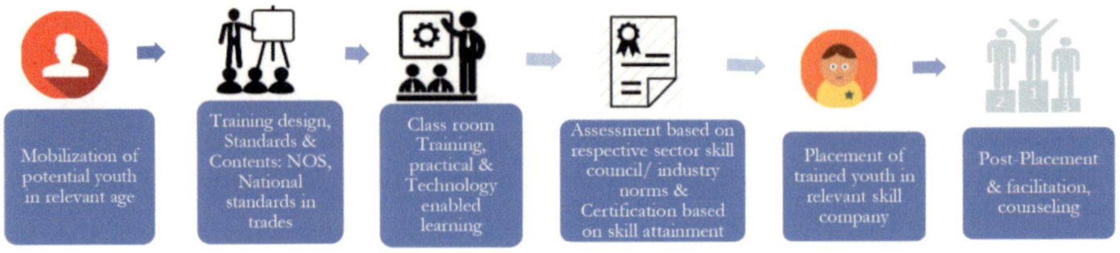

Figure 4.
Learner engagement model. Source: [23].

are conveniently packaged in learning skills with skill attainment and assessment, which are modular based. Upon completion of the entire learning hours, the learner is equipped to deliver services in the respective trade and competent to be an expert in the learned domain areas.

Figure 4 illustrates the typical vocational service provider engaged in training in skill domain areas. The typical learner engagement model is illustrated.

4.1 Sequential steps about the placement-linked employment training model (PLET)

4.1.1 Mobilization

Youth are mobilized through local resources (representatives—panchayat heads, village heads, local contact point such as postman, bank representatives with some kind of incentive schemes), and information is also collected using technology-enabled solutions—mobile solutions, and identified youth are invited to the nearest skill centre for counselling and enrolment in the skill programmes.

4.1.2 Enrolment and Skill Training

Upon completion of skill assessment and enrolment in programmes of the skill centre, the trainings are held as per the standards. The skill programmes follow the prescribed National Occupational Standards (NOS) and the National Skill Qualification Framework (NSQF). The skill centres also have unique training delivery methodology addressing the needs of learners, a unique approach of Module Centred Learning Architecture (MCLA), where the trainings that are classroom combined with video-based training delivery are followed. This method is also an ideal, where teachers and trainers are remotely available for continued interaction online with learners.

4.1.3 Vocational Pedagogy

The skill training is a mix of classroom teaching, using of video-based content, mentoring remotely, case study-based learning—practical training, on-job training/industry-based training.

4.1.4 Assessment

An assessment to the training is also included, which includes domain skills, practical knowledge & soft skills, competency based. The grades for the skill acquired are based on the learner's competency, which are also assessed by the employers.

4.1.5 Post-training and review

Post the training and assessment learners, at this stage trained youth is placed with the employer based on the skills certificate obtained from the employer database obtained already for placement. Where needed, the trained youth are mentored and counselled during the career to meet the individual needs and organizational objectives.

4.2 Technology in vocational education training or ICT in vocational education and training

Technology refers to use of computer-aided learning and the innovative use of information communication technology in order to achieve the learning objective in technician and vocational education in the time and scale. They are most often as virtual teachers in guiding the learners when enabled with virtual or remote teachers for a large group or learner community.

A variety of technology-enabled solutions are developed by vocational and skill training providers to meet the larger objectives of flexible learning and lifelong learning. The type of technology varied with the type of skills and vocational trades offered by the service providers. They can be broadly classified and termed as follows:

i. Multimedia technology for interactive learning

ii. Online learning portals

iii. Virtual labs

Each of the aforesaid can be defined by means of their use, application and relevance for vocational and skill training. The multimedia technology for interactive learning is multi-modular-based skill training resources wherein trainer can use theses as teaching aids and after the classroom delivery, learners can use them as self-learning module. These can be used based on the learners' skills attainment, capability to learn and capture the skill levels and advance in the learning attainment. The teacher/trainer can use the multimedia content to augment this as teaching notes and combine the virtually additional resources to given learners additional information, techniques in variety of modules. This serves multiple functional pedagogical techniques.

At the same time, online learning portal is a collection of ICT-based vocational and skill contents, already available in the Web site or a pre-recorded content and depending on the learner capacity and requirements, which can be augmented for teaching and learning. However, this allows flexibility to use wide content, which can be sourced from the Web resources across a variety of areas and geographies. The learner has the option of choice for a specific level of learning speed and accordingly can access the resources and be able to combine the best resources available in the Web-based resources. This shall have seamless access to resources available on the web; however, the limitation include they need to be customized as per the skill level, type of trades, focus to skill attainment and too much of flexibility in the assessment framework.

The virtual labs are unique Web-enabled practical handling technologies wherein students will be guided through specific instructions to undertake hands-on training, practice with virtual experimentation and clearly defined outcome-based learning. This will work within the given framework of experimental theme and programme. These are also most suited to advanced experiential learning and experimentation.

4.3 Augmented reality (or) virtual reality in vocational education training

The learning technology for vocational and technical education has grown to a more technologically development methods, by application of real-time learning using virtual experiential learning. This is made using application of more augmented reality, wherein the learner applies using joystick or mouse to experience the real life-learning virtually.

The author in a project funded by the United Kingdom India Education Research Initiative (UKIERI), a bilateral funding for education, TVET and skill development initiatives developed a model for learning and training the vocational trainers; that is, trade instructors and faculty of TVET providers (all beneficiaries were from the government-supported institutes and skill centres) in the welding technology and painting technology, using augmented extended reality models.

The extracts of the use of augmented reality learning contents are presented and discussed briefly, as a case study model. A survey of vocational and technical education trainers were interviewed regarding the use of modern teaching technologies. It was found that the majority of the trainers had access to the resource, however, could not find them relevant in the context of skill attainment; it was only after practical demonstration in the workshop, lab, the learners are able to respond, react and provide ideas on the learnings. Most of the vocational training providers in the government space also have lack of access to modern lab and demonstration equipment and devices. These vocational providers have limited access to updated/latest learning resources. Almost all vocational providers funded by the government at the local, regional and national levels have computing facilities with access to Internet sources. This allows trainers and learners the access to the world of Web resources. However, the access to such resources depends on the type of skill and vocational programmes available in the centre, extent of knowledge and skills of trainers for accessing such contents, etc.

It is in the context, the UKIERI-funded programme attempted to provide knowledge and information regarding the use of augmented extended reality in technical education and training relevant to certificate-level trades covering selected states in India. The extracts as figures of the extended virtual reality as an example for two trades covering welding technology and painting are shown under.

The present learning technology emulated by the several of the vocational content development service providers includes application of artificial intelligence-based extended virtual reality using machine learning processes. The process of artificial intelligence allows to have choice of certain commands that apply to the actual demonstration of experiments, lab techniques, etc. Hence, they allow learning with proper logical approach and sequencing. This allows learner a choice and level of learning in the vocational education training sector (**Figure 5**).

A typical model for a welding technology using augmented reality demonstrated during the training is illustrated in **Figure 6**. It may be seen the trainer using virtual models and joysticks for the welding trade.

In general, Artificial Intelligence and Machine Learning unit in a simulator is fed with a large amount of historical data dexterity information, technical field values, for example voltage, current flow rate for any typical engineering sector, and the unit is trained using supervised learning technique, past inputs and corresponding outputs used to create predictions. For example, in the case of welding training, for a given set of input hand movement, current, voltage and plate thickness and related demonstration parameters would be the welded joint defective, the defects

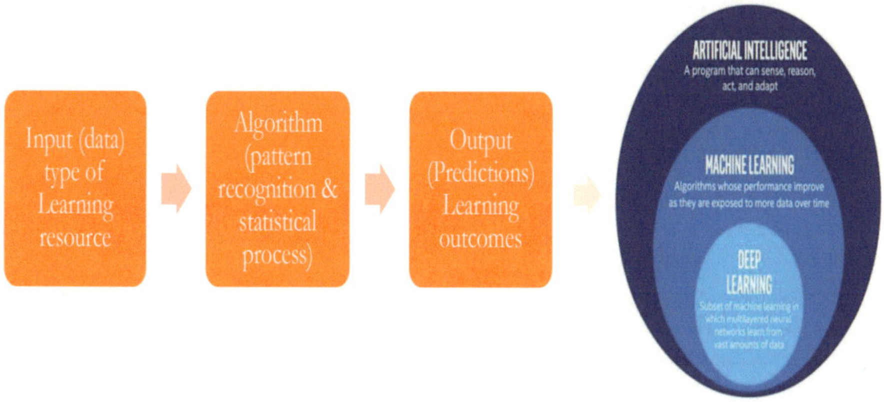

Figure 5.
Application of AR-VR with AI in TVET learning contents (project partnered with UKIERI, the Ministry of Skill Development and Entrepreneurship, 2019). Source: Skillveri Training Solutions Private Limited, Chennai, India, 2020.

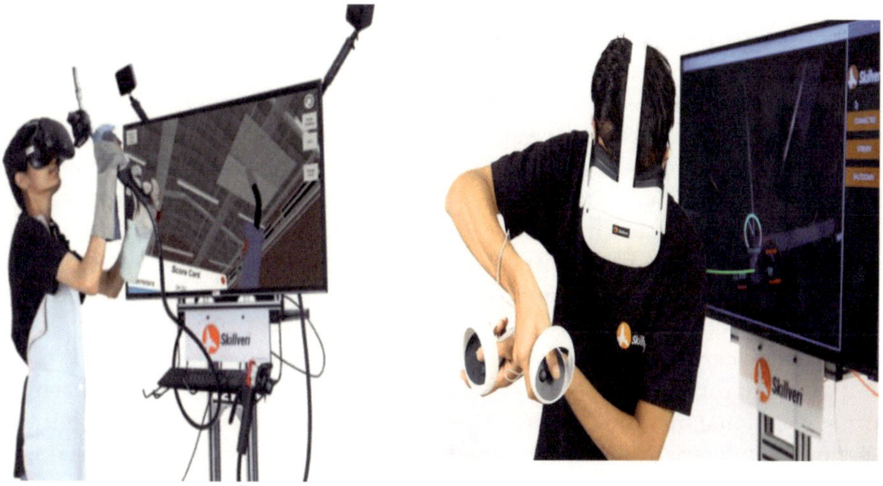

Figure 6.
Examples of extended virtual reality in welding technology relevant to skill qualifications for skill qualification framework levels 4 and 5. Source: Skillveri Training Solutions Private Limited, Chennai, India, 2020.

reveal the demo and be able to predict and show the learners/trainers. This interface creates enhanced learning environment by developing links between the inputs and outputs and the impact of each input parameters on output.

Mixed reality is yet another new concept, where learners have the option to learn from the mixture of real objects and real working experiments with virtual models. The key aspect that is more relevant and appropriate is included for the engineering and manufacturing sectors. The learner thinking skills along with the real objects help them to conceptualize what shall work better in the given context and situation.

5. Conclusions and learnings

With a wider option for the vocational and skill development, learners' aspirants and trainers, multiple options exist with the development of technology and customized development of such technology-enabled learning resources. Many a times, the modules are specific to discipline and skill trades. The trainers and learners can have choice from classroom lectures using books, textbooks, skill manuals/guides, computer-aided learning resources, hand-held devices and a wider range of electronic appliance such as tablets, smartphones, short clippings from multimedia resources clubbed with any Web-enabled contents. There is now wide choice of knowledge resources, great deal of information, and hence, adopting to appropriate methods and application of type of resource is dependent on the learners' capability, type of discipline/trade and skill relevant to specific need and demand.

The advancement of learning in a virtual environment shall be largely based on the application of the augmented reality, virtual reality or extended virtual reality, which shall be one of the most advanced developments in vocational and skills training. The application of artificial intelligence will help suit the learners' mode in making the right choice of the content relevant to specific trade and learners' ability towards skills attainment.

6. Definitions

TVET means Technical Vocational Education and Training, which is understood as comprising education, training and skills development relating to a wide range of occupational fields, production, services and livelihoods. TVET as part of lifelong learning can take place at secondary, post-secondary and tertiary levels and includes work-based learning and continuing training and professional development, which may lead to learning-acquired certification/qualifications [24]; e-learning means learning done by learners using computers and computer-based programmes and TVET refers to computer-based learning of vocational- and skill-based contents, with the aim to meet the learners' learning needs and requirements, usually relating to additional information and clarification of complex issues; virtual learning environment means education technology using a Web-based platform for the digital aspects of courses of study, usually within learning centres/vocational institute, and they are a pool of learning resources, activities, assessments, reference points, combined with interactions with course structure, and are provided for different stages of assessment [25]; augmented learning technology means is an on-demand learning technique, which adapts to the learner requirements, by providing remediation on-demand, learners can gain greater understanding of a module, topic, etc. These technologies incorporate rich media that the learner experience an adaptive learning experience based on the current context [25].

Acknowledgements

The author is thankful to the grants made available for a project through the UKIERI for 'Train the Trainers of Industrial Training Institutes (ITIs) covering select states in India' and piloting the use of Artificial Intelligence based Extended Augmented Reality welding and painting technology module and sharing the part research in this chapter. Pictures and model of Skillveri Training Solutions Private Limited, Chennai, are also suitably acknowledged.

Author details

Balakrishnan Chandrasekar
Homi Bhabha National Institute, A Deemed to be University, Mumbai, India

*Address all correspondence to: drchandrashekar26@gmail.com

References

[1] Wood's Despatch. Report of the Committee on Education. 1854

[2] Government of India. Report of the Indian Education Commission. Calcutta: The Superintendent of Government Printing; 1882

[3] Sapru Committee Report. Report on Indian Education. 1934

[4] Abbot A, Wood SH. Report on Vocational Education in India. Delhi: The Manager of Publications; 1937

[5] Basic National Education. Report of the Zakir Hussain Committee and Detailed Syllabus with a Forward by Mahatma Gandhi. Wardha: The Secretary, Hindustani Salami Sangh; 1938

[6] Central Advisory Board of Education. Report on Post-War Educational Development in India. New Delhi: Government of India Press; 1944

[7] Government of India. The Report of the University Education Commission 1948-49, The Ministry of Education. Delhi: The Manager Publications; 1949

[8] Government of India. Report of the Secondary Education Commission, The Ministry of Education 1952-53. 1953

[9] Central Advisory Board of Education. Vocational Education for Agriculture and Industry, 33rd Meeting. 1967

[10] Government of India. National Policy on Education 1968. New Delhi: The Government of India Press; 1968

[11] Adiseshiah SM. Vocationalization of Higher Secondary Education and +2 Committee 1978. National Review Committee of the Union Ministry of Education, Government of India; 1977

[12] Chandrasekar B et al. Review of Technical Vocational Education and Training System and Skill Development System: An Analysis of the Structure of the TVET Related Institutions in India. Issue no. 13. 2009. Available from: www.tvet-online.asia

[13] Government of India, the Kothari Commission. Report on Education Commission 1964-66, Education and Development. New Delhi: Government of India Press; 1965

[14] Available from: https://msde.gov.in/

[15] The World Bank. Skill Development in India: The Vocational Education and Training System. The World Bank document. 2007

[16] Central Advisory Board of Education. Report on CABE Committee on Education 2008, Ministry of Human Resource Development; 2008. Available from: www.mhrd.gov.in

[17] Gazette Notification dated 27 December 2013. Government of India

[18] Okoro OM. Principles and Methods in Vocational Education and Teacher Training. Nsukka University Trust Publisher; 1999

[19] Winer RK. Rung by the health career ladder. American Vocational Journal. 2000;**48**(7):18-27

[20] UNESCO-UNIVOC. 2017. Available from: www.unesco.org

[21] Diegmann P, Schmidt-Kraepelin M, van den Eynden S, Basten D. Benefits of Augmented Reality in Educational Environments—A Systematic Literature Review. 2015. Available from: http://aisel.aisnet.org/wi2015

[22] The World Bank Development Indicators. 2019. Availablem from: www.worldbank.org

[23] Chandrasekar B. How the Private Sector Develops Skills. India: UNDP Istanbul International Centre for Private Sector Development, UNDP; 2015

[24] UNESCO-GC. 2015 Available from: https://unesdoc.unesco.org/

[25] Available from: wikepedia.org

Section 2

Augmented Reality and Engineering

Chapter 6

Intelligent VR-AR for Natural Disasters Management

Silvia García, Paulina Trejo and Alberto García

Abstract

Considering the significance of improving natural disasters emergency management and recognizing that catastrophe scenes are almost impossible to reconstruct in real life, forcing persons to experience real hazards violates both law and morality, in this research is presented an engine for Virtual Reality/Augmented Reality (VR/AR) that works enhancing human capacities for prevention, response and recovery of natural phenomena effects. The selected novel techniques have very advantageous qualities to overcome the inconveniences detected in the most recent seismic devastating experience in Mexico City, the Sept 19th, 2017, earthquake M7.2: total collapse of more than 230 buildings, partial fall of 7 000 houses, 370 people were killed, and over 6,000 were injured. VR and AR provide researchers, government authorities and rescue teams with tools for recreating the emergencies entirely through computer-generated signals of sight, sound, and touch, when using VR, and overlays of sensory signals for experiences a rich juxtaposition of virtual and real worlds simultaneously, when AR is applied. The gap between knowledge and action is filled with visual, aural, and kinesthetic immersive experiences that poses a possibility to attend to the population in danger in a deeply efficient way, never experimented before.

Keywords: virtual reality, augmented reality, mixed reality, artificial intelligence, Mexico City, September 19th 2017 earthquake, Seismic phenomena, management natural disasters

1. Introduction

With the increased prevalence of smartphones and popularity of artificial intelligence (AI), substantial research and development have been made in pursuit of more cost-effective and high-performance sensor technologies [1, 2] and graphical processing units, which have led to the production of affordable virtual and augmented reality devices [3]. These developments allowed researchers in many fields (e.g., astronomy, psychology, medicine) to create controlled virtual environments that permit users to interact with digitally generated stimuli [4–6]. Virtual Reality (VR) and Augmented Reality (AR) are presented here as present and future alternatives for the efficient training of human resources, not only in the industrial field, but in every aspect of real world, of real life. The virtual recreation which objective is to transport the user to fully digitized and interactive environments, has allowed for the simulation of real processes and situations, an adequate way to reduce training times, prevent errors, or even improve the quality of products and actions. In the field of environmental sciences and natural disaster management

(NDM), public, scientists, decision-makers, and professionals can benefit from virtual and augmented reality applications to simulate various and complex scenarios, constituting a realistic and safe workspace for repetition, precise measurements, and improved knowledge [7, 8].

In this research it is presented how VR and AR are applied to digitally recreate the real-life setting when an earthquake hit. In the preparedness before disasters happen, the simulated scenarios allow the user to interact with buildings, houses, foundations, streets, and buried pipes, on the one hand, and with soil strata, rock basement, ground motions, fissures, and cracks, on the other. Tricking the human perceptual system into believing that is being part of the virtual city, it is guaranteed complete focus on the effects of the seismic shaking.

By the seamless blending between a real environment and computer-generated virtual objects, the resulting mixture supplements the natural environment (the phenomena, the infrastructure, and soils responses) rather than replacing it with simplified conceptions. The virtual objects incrusted in the reality display information about structures characteristics and soils properties that the users cannot directly detect with their own senses. This information conveyed by the virtual objects helps a user perform better NDM-tasks. The concepts of VR/AR activated are simulation, interaction, artificiality, immersion, telepresence, full-body immersion, and network communication. The input information is compiled from the September 19th, 2021, earthquake files (accelerograms and geotechnical zonation where the monitoring station is placed), records of large settlements, fissures, and cracks as well as total and partial collapses of buildings and damage to buried pipes. The devastating earthquake of 2017 provided the opportunity to evaluate how citizens and the government prepare and respond to emergencies due to natural phenomena, particularly the seismic phenomenon, to generate the necessary improvements at each stage of natural disaster management.

2. Why natural disasters management must be supported by VR/AR

Built environments, which refer to all physical environments constructed for human habitation and activities [9], are constantly exposed to risk from various natural and manmade disasters, such as fires, tropical cyclones (wind and storm surge), earthquakes, tsunamis, floods, and terrorist attacks, which pose a significant threat to human beings. Considering the severity of natural hazards, that become disasters when people's lives and livelihoods are destroyed, an appropriate NDM is crucial to reduce the harmful effect of the phenomena and to facilitate a prompt reestablishment of the normal life after the emergencies [10]. When the threatened scenarios are large cities or strategic sites for nations, it is essential to have efficient and timely management programs that protect lives and properties of the inhabitants but also that quickly restore the productive and commercial capacities of population centers [11]. NDM covers i) risk prevention, ii) crisis preparedness (specific training), iii) emergency response (rescue), and iv) catastrophe recovery (reinstatement of services and lifelines) [12]. The key point in these actions is the preparation facing of the emergency so the researchers and professionals must work to improve the understanding of the situations, processes behind and analysis methods, anticipating the adverse possible situations, their worst evolution, and forecasting how they arrive to critical states.

The NDM process to save human lives, divided in pre-, during-, and post-disaster actions, still have a gap between knowledge and *intelligent* and effective procedures [13]. The impossibility to prepare and train under real-life conditions, limits the learning experience of first-responders, civilians, and city-planners. During disasters, communication, and visualization of possible pathways to recovery (since danger could be permanent and/or resources are scattered) is fundamental. In this sense, VR and AR technologies cannot be ignored anymore: they have grown exponentially over multiple markets and is projected to have a worldwide revenue of USD 40 billion by 2024 [14, 15]. The examples of successful applications are varied and motivating: simulation of floods, ground motions due to earthquakes, wildfires and hurricanes, around the training, the monitoring, the modeling and the early warning [16–28]. The virtual and augmented scenarios, most of the times, are developed as smart systems (based on artificial intelligence) to improve situations prediction [29–34].

Between the various studies the use of smartphones is indispensable. The potential and benefits of using AR to access and display disaster data in its geospatial context [35–37] is normally developed through a mobile application, for example to determine the best route for evacuation from collapsed or burned areas. The Whistland system [38] is an example of retrieving crowdsourced disaster-related information from real-time Twitter feed for display as an AR overlay to the smartphone camera. Mirauda et al. [39] developed a mobile application for connecting measurements from hydrological sensors with an integrated forecasting model. Fedorov et al. [40] presented a unique approach to utilize computer vision techniques to identify mountain silhouettes and compare the extracted information to available DEM (digital elevation model) data to provide useful information (e.g., peak name, height, lithology, associated risks, etc.).

About AR/VR solutions the best models are those developed for visualizing data and analyzing information in-situ in smooth transitions between virtual and augmented environments [41]. Ready et al. [42] presented a virtual reality application for HTC Vive that recreates a 3D model of terrain and buildings (geographic context in Japan) to interact with various data resources, mainly to access hydrological time-series in an easily interpretable way for disaster management. Haynes et al. [43] developed a mobile application to visualize potential floods through integration of real-time recordings (e.g., water level, soil moisture, and humidity), being extremely important that stakeholders can observe situations from the application and make decisions consistent with what happens in the field. Macchione et al. [44] proposed a virtual environment (open-source 3D graphics app, i.e., Blender) to recreate an urban environment (e.g., buildings, streams, roads, levees, textures) to simulate hydraulic dynamics during different flood scenarios.

3. VR: AR basics

Virtual reality (VR) is a technology with which digital environments can be built for the immersion of participants. In these environments, users have the possibility of interacting with the environment and with each other in deeply realistic ways [45]. The ultimate goal of VR is for participants to experience total immersion and prevent, in every way possible, from perceiving stimuli from the outside (real) world [45]. However, VR also involves non-immersive and semi-immersive settings (i.e., when the experience is in a desktop) [46]. On the other hand, Augmented reality (AR) concentrates on the superposition of virtual or digital elements on real world scenes. The participants can "view" the real world, but it is composited with digital items [47]. AR expands the real-world rather than constructing a virtual one. The computer-generated material improves understanding or understanding of what is happening in reality [48]. Mixed Reality (MR) is a hybrid technology where virtual objects are merged into a 3D- atmosphere or real objects are positioned into a virtual creation (**Figure 1**).

Figure 1.
Application example for the Federal Electricity Commission in Mexico, the team of engineers works a) in a geotechnical laboratory with an application that allows them to learn more about the specimens being analyzed, share the results with supervisors (in real time) and generate the log files using the lenses; b) in discussing projects in a more immersive way in which all sources of information can be displayed to each team member can understand and discuss without bias.

Another definition of VR is a specific collection of technologies (headset, gloves and walker) that allow participants to feel that they belong to a digitally created world through high interaction between their senses and the artificial environment [49–54]. The 3D simulation (width, height, and depth) of a real or made-up situation can be experienced visually in real-time motion (or the closest thing to this) and supported by sounds and tactile stimulus, or the necessary feedback to recreate an integral experience. With VR, users can envisage, operate, and relate multifaceted data with surprising ease [55]. Therefore, VR refers to an immersive, interactive, multi-sensory, viewer-centered, 3D computer-generated environment and the combination of technologies required to build such an environment [56, 57]. The stereoscopic ambience enables the observer experiences in deeply immersing scenes. Exploiting the human brain responses, VR dilutes the boundaries between persons and computers. Because of our ability to see the environment three-dimensionally (stereoscopic vision) VR can create right and left eyes images of an object or a scene and the observer's brain integrates these stimuli from the presented perspectives to generate a whole sensitivity of a space. The virtuality means that an illusion is created about screen objects beyond the information on 2D displays in monitors. In VR, as in other technologies (2D-CAD even the 3D version), it is tried viewers notice distance and spatial interactions, but more convincingly and precisely.

3.1 Components of VR/AR

An augmented reality system, from a hardware perspective, consists of a sensor(s), processor(s), and display(s). AR systems superimpose computer graphics imagery on the real world, and, for this blending, the observer's positions must be known (with extreme precision), stored, and related with the positions and geometry of the items that are to be projected on the real environment. From a general perspective, it is quite easy to draw the three-dimensional images that you want to superimpose on what the user observes in the real world. The task to be solved is to fully define a correct perspective while defining as accurately as possible the position of the observer's eye. To offer a truly useful experience, portable and easy-to-use optical systems must be considered.

The AR processor coordinates and analyzes sensor inputs, stores, and retrieves data, carries out the tasks of the application program, and generates the appropriate signals to display. Computing systems for augmented reality can range in complexity from simple handheld devices such as smartphones and tablets to laptops, desktop computers, and workstation class machines all the way through powerful distributed systems. The scene must be updated smoothly and at a rate that the participant in the experience perceives as a constant stream of information. AR applications must sustain a frame rate of at least 15—preferably more— frames per second for the participant to perceive the display as continuous. Displays that are simulating the feel of a solid object must be updated about 1000 times per second or else the object will feel "mushy." To superimpose information (graphics or data) stored in a computer (control center) on the actual environment, a device called a beam splitter is typically used. This divider does not divide but combines the images of the real environment with those placed in the monitor environment. As a result, the viewer is presented with a double exposure photograph. Because the optics are typically fixed, in AR systems there is only one depth at which both the computer-generated imagery and the real-world imagery are in focus. If real-world and virtual-world scenes are both in focus, it will be easier to perceive them simultaneously.

By the other hand, the components necessary for building and experiencing VR are divided into two main components, the hardware and the software mechanisms. The hardware components are composed of computer workstation, sensory displays, process acceleration cards, tracking system and input devices. The workstation is an object (computer or microcomputer) that is intended for technical or scientific work. These objects are typically used by one person at a time, but are connected to a network to facilitate multi-user operations. Sensory screens are the artifacts that are responsible for displaying virtual environments. These displays can be as basic as common computer display units to head-mounted displays (HMDs) with headphone mounts for a 3D viewing and audio experience.

When over-the-head (helmet) displays are used, the presentation of the images is given right in front of the viewer's eyes. With the helmet's sensors (the orientation ones) the artificial segments are controlled that make the user experience a complete virtual environment. In most cases, a set of optical lens and mirrors are used to enlarge the view to fill the field of view and to direct the scene to the eyes [58]. It is fundamental to have process acceleration cards to update the display with new sensory information and the tracking system (mechanical, electromagnetic, ultrasonic, and infrared trackers) that follows the position and orientation of a user in the virtual environment.

There are also VR software key components: 3D modeling software, 2D graphics software, digital sound editing software and VR simulation software. 3D modeling software is used in constructing the geometry of the objects in a virtual world and

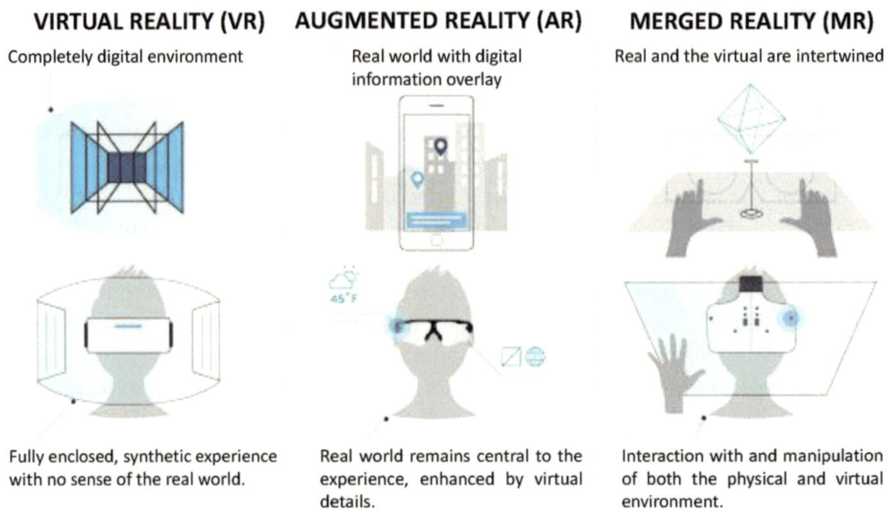

Figure 2.
Reality roadmap according to Intel®, which prefers merged reality to mixed reality; the term "mixed reality" has been thrown about a lot as of late but pinning down a precise definition has proven elusive.

specifies the visual properties of these objects, and 2D graphics software permits to manipulate texture to be applied to the objects which enhance their visual details. With the digital sound editing software is mixed and edited sounds that objects make within the virtual environment and with the VR simulation software the components are put together. It is used to program how these objects behave and set the rules that the virtual world follows.

In this research is presented a mechanism of Mixed Reality (MR), which combines Virtual Reality and Augmented Reality, called also 'Hybrid Reality' or 'Extended Reality'. MR [59] is understood as a proposal that can be placed anywhere between digital environments that belong to complete virtuality up to the absolute perception of the real environment. MR works with concepts and objectives of AR and VR (**Figure 2**). The use of helmets or glasses allows the user to enter a digital world with all the information that could be useful. The continuity between reality and virtuality is the basis for the interaction between objects in the physical world with items in the virtual world, which will finally be the future of these technologies.

4. Application for earthquakes

Modeling objects and the building structure in the virtual world is the first step for environment construction. In this research the employed techniques entail software for modeling/constructing bi- and three- dimensional objects and backgrounds. Once the 3D modeling of the objective scene is built, the model brings in the VR/AR mechanisms (in this research is used Unity3D-Unreal), in this way the static model is converted into a synergistic environment. Based on the experiences modeling urban infrastructure, these alternatives are appropriate for metropolis projects. Then the virtual contents (movements, gestures, and commands) were established. Physics engines were sometimes used to assist the simulation of emergencies, i.e., for a building shaking according with the accelerations registered during an earthquake, in each virtual room the movements of the room must be experienced, but in addition to the objects that could also move and even fall.

An avatar is used to manage the commands that will make the experience in the virtual environment one with dedicated goals. The avatar is the representation of the person, and it reflects the response behavior (of him and/or the participants) to the stimuli of the virtual world. For an avatar to move and act, a device such as a mouse, keyboard, pad, and telephone (in the 2D case) and gamepad and joystick (for the 3D offer) is used. The AR phase contains, in addition to the possibilities of movements and executing commands, data and information from the real world that could be useful for the designed task. Cameras and sensors are widely used and combined with the image recognition function of an AR engine. HMDs (glasses) provides the ergonomic solution for a virtual environment task using direct vision. Users can experience high-level immersion and better react as if they were in the real world.

The September 19th, 2017, a M 7.1 earthquake hit Mexico City causing 370 people were killed, over 6,000 were injured, 230 totally collapsed buildings and more than 7 000 small houses were damaged. The situation required the intervention of hundreds of geotechnics and structural specialists to qualify conditions and to permit people to return to their homes (**Figure 3**). Exactly 32 years before (Sept 19th, 1985, more than 10,000 deaths, 30,000 destroyed buildings, and 68,000 injured people) a M8.1 event had showed the fact that the city is built on the unstable and sinking ground of a dried-out lakebed and this promoted a better geo-zonification of the risk.

In 1985 the functioning of the city was failed for months, the reconstruction took years and led many to relocate from the most affected areas to the city's outskirts in search of the safety of the bedrock. The districts affected in 1985 and 2017 events were quite different and, because of this, the type of damage and the technical needs to manage rescue activities were also distinctive. These experiences have undoubtedly shown that for designing an efficient post-earthquake relief plan, it must be recognized that Mexico City is particularly exposed because of its huge population and strategic importance for the country and that this vulnerability has grown the last years due to the expansion of the urban settlements in risky areas (nodules of extreme poverty), the environmental devastation, the deterioration

Figure 3.
The Mexican earthquake from the 19 September 2017 led to significant building damage in the capital Mexico City and the states of Morelos and Puebla. The damage data in houses/buildings and soils characteristics highlights the correlation between damage drivers that little has been studied or they have not been fully understood.

of life levels, the economic activities concentration that require the transport of substances (water, potable and used, gas and hydrocarbons) by underground infrastructure, and the growing complexity of transportation process.

Despite the best intentions and the enormous efforts of the governments that have attended these emergencies, the situations dangerously evolved and the period in which the city was detained, and the population subjected to chaos, spread out for months (or even years in 1985) damaging, primarily, the poorer and fragile (socially) population centers. Hereby, there is an increasing need for a holistic and more efficient natural disaster management.

The application that is explained in the following is VR-AR engine to train and to administrate the three phases of this kind of NDM: preparedness, response, and recovery. This VR instrument has the capability bridge the gap between knowledge and action with the necessary velocity and efficiency when disasters happen. Exploiting the scenarios reconstructed, the first-responders, civilians, and city-planners can be prepared to work under a series of conditions built in virtual/augmented-life, expanding the learning and response experiences. The heterogeneity of soils properties (and responses), earthquakes damage intensities, and the actions effectiveness (professional teams and their capacities) is displayed in the digital metropolis based on the results from neural topologies that predict the spatial variability of risk components (exposure and vulnerability).

4.1 Pre-emergency preparedness

For the preparation of engineers and specialists to perform i. state declaration (for soils and structures), ii. routes recognition (safest and more efficient), and iii. Provision of resources, it was necessary to construct specific VR-AR tools. The training under real-life conditions goes in two directions: geo-situations and structures. Simulations are about first-responders' reactions for specific scenarios: 1. extreme vulnerable communities, 2. structural pathologies (residential homes and multi-family buildings) and 3. buried infrastructure and 4. soil masses.

The immersive experience of an earthquake (during the shaking and/or immediately after it has happened) is simulated in locations where fragile infrastructure coupled with danger zones (susceptible soils). In **Figure 4** is shown how is recreated the structural masonry commonly used in poverty nodes in Mexico City. The structuration refers to the practice of using masonry, brick, or stone, as mass self-supporting. It is one of the oldest building methodologies, and by far the most resilient, however when self-construction (the inhabitants develop the structuring without the support of engineers or architects) does not follow basic rules for the placement of vertical and horizontal reinforcement elements (columns and beams) and the mezzanine and ceiling slabs are extremely light (even without steel reinforcement), the response of these units to ground movements is very unfavorable.

The VR-AR tool (principally the VR box) drives the user to detect first the kind of cracks, fissures or other pathologies presents in a house and to relate them with a classifying guide (**Figure 5**). After this, the observer must look for overgrown trees and shrubs, cracked drains, leaking rainwater goods, as some of the things that may lead to structural problems but that are not related to the seismic inputs.

Part of this investigation, but still in a preliminary stage, is the issue of the entry and exit routes of the emergency teams. The evacuation of threatened communities and assistance to those potentially blocked by the effects of an earthquake is a complex and vital issue. Transportation system is conceived as the role that sustain the economic and social well-being of the communities so disaster or extreme hazard such as earthquake has a major impact on the resilience of the communities. In Mexico City suburbs, road infrastructure is linked to many factors such as users,

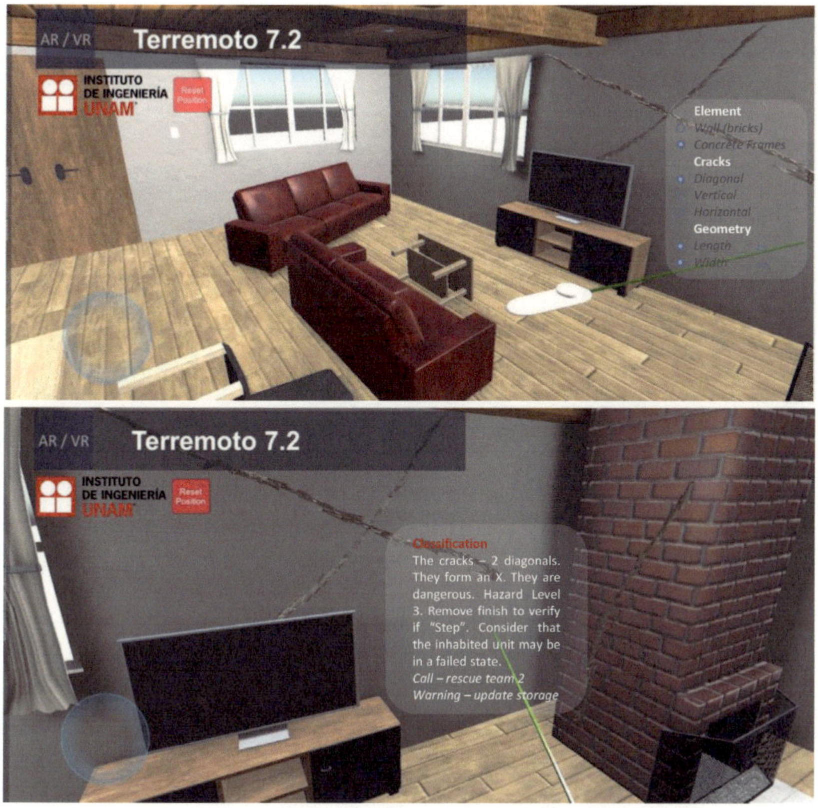

Figure 4.
A training room: after an earthquake M8, subduction zone, site Lake zone, failure conditions for a concrete frame system (1 floor). Analyzing cracks in walls and columns the user is being trained to correct classify the damage level after an extreme earthquake.

climate, economic level, material, topography, and periodic maintenance, therefore, part of the real-time analysis for the planning of entry and exit routes must be the superposition of layers on information from massive sources (traffic control systems) and the spatial variation of these susceptibilities. A comprehensive review of social infrastructure (hospitals, schools, recreation centers, markets, department stores, fire and police stations, etc.) must be done to detect temporal routes options as part of the adaptive routing solution.

4.2 Response during the emergency

The Response during Emergency refers to the actions that people may take, in the case of earthquakes, immediately after the ground movement has ended. The emergency period, when the events are extreme, can be extended depending on the size of the heavily damaged areas or in which the collapses have occurred in public places with large concentrations of people. The AR box was developed to assist technical crews which are organized according to the risk levels at each site. The entrance to housing units is categorized according to their structure and size: types of walls, types of columns, types of slabs, types of foundations and, on the other hand, single-family houses, multi-family houses, light and small buildings, large buildings.

In this stage of NDM, the personal is in real danger so it is mandatory that they have a support outside the damaged site: an automated or human control center. The superposition of information about the structures and materials permits rescue teams to take immediately decisions about evacuations or calls for additional help

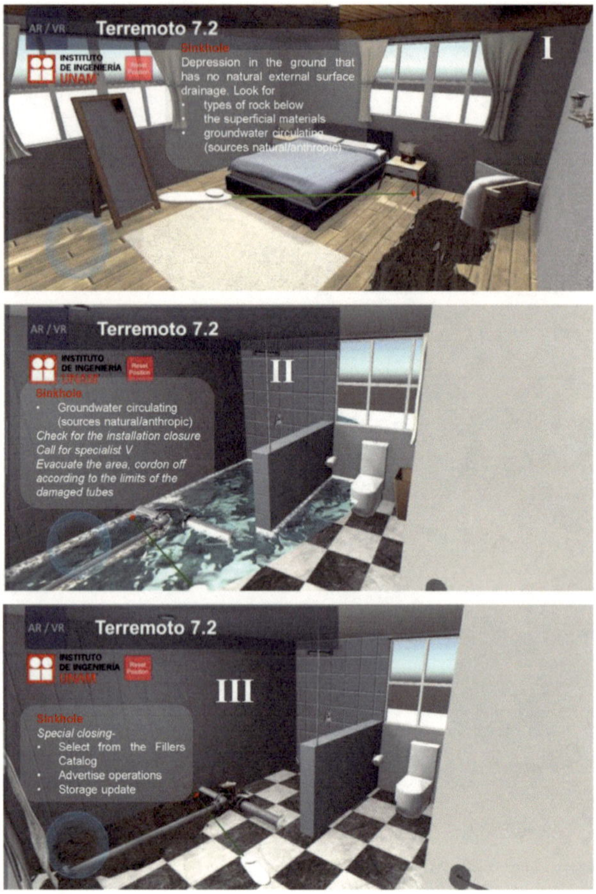

Figure 5.
A training room: after an earthquake M8, a sinkhole has formed, and it is necessary to determine the sources that erode/dissolve as well as call the specific help team. The repair conditions are stored in the event DataMart.

(**Figure 6**). Also, it is important that the field teams are provided with a dedicated storage for the found conditions. Once the *in-situ* situations are loaded, they are integrated to a set for specific analysis (feed forward neural network) that defines the risk level and spatially categorize the geographical situation. In this way, state maps are built in real time. Additional maps can be displayed using the stored information, for example by factor, by action demanded (posterior attention, evacuation, or human rescue) and by supply (call for requirement of special equipment/machines or other inspection/rescue teams) (**Figure 7**).

In addition, there is a section of AR toolbox that is exclusively dedicated to the attention of leaks. Water and gas, the latter considered a priority due to the secondary effect of the explosions, are attended by specialized professionals that work in accordance with specific regulated policies (**Figure 8**).

In some zones of the metropolis the vulnerability of the soils to the arrival of seismic waves is very high. The most superficial layers in these areas suffer cracking and subsidence processes among the most alarming scales in the world (**Figure 9**). The periods of drought and torrential rains aggravate the susceptibility to the collapse/cracking. This situation maintains small buildings and buried facilities in a very susceptible state, making them prone to be more affected when an extreme earthquake hit. The coincidence between construction deficiencies (poor technical conceptions) and degradation of materials (highly deformed, cracked and collapsed soils), when the seismic waves arrive, constitutes a challenging scenario.

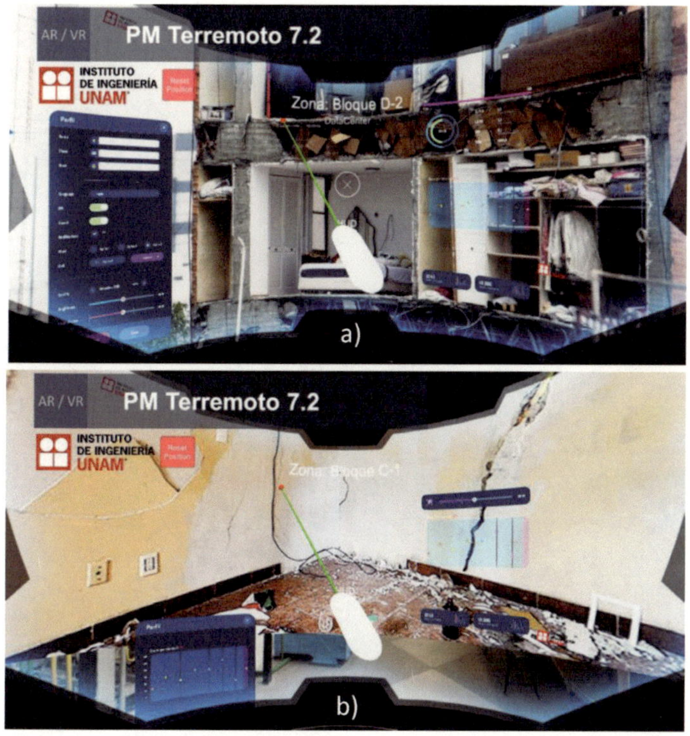

Figure 6.
Immediately after the earthquake has ended, the trained teams go out into the field and begin to verify the conditions through the lenses that are communicated with the control room, a) since street, an engineer is checking a multi-level building that has lost the walls on one side of its perimeter, b) inside a 1-story house, a user checks the cracking, catalogs it and stores it.

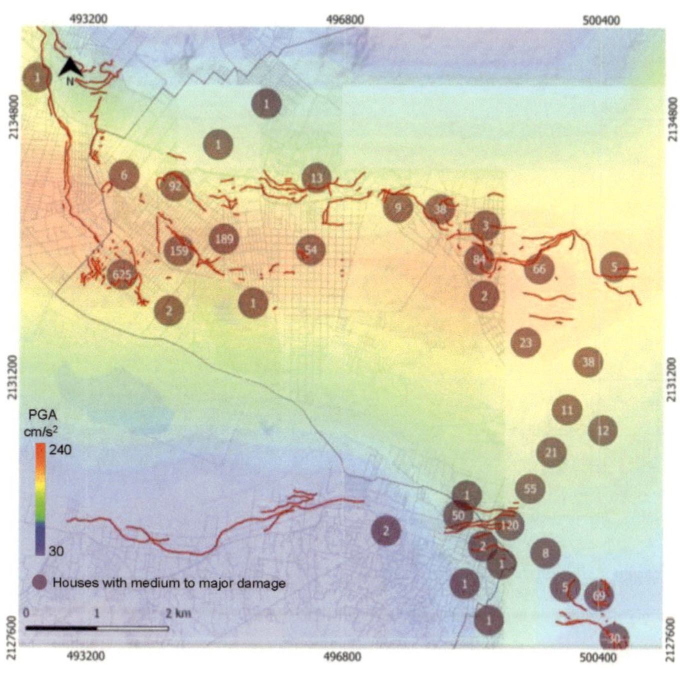

Figure 7.
Example of the integral maps: peak ground accelerations PGA superposed on the number of damaged houses (1 to 3 floors) -red circles- in a small region of the Mayor's Office of Tláhuac, southeast of Mexico City.

Figure 8.
The status review of the pipelines that transport dangerous substances (in this case gas) is essential to qualify the inhabited regions as safe. The application created for one of the largest companies in the country is shown. The user has the information about the arrangements in the field as well as the readings that must be verified on the special monitors. At all times, the user can be assisted by his human colleagues in the control room.

Figure 9.
In the southeastern of Mexico City, where the poorest neighborhoods of the metropolis are concentrated, during the 2017 earthquake manifestations such as cracks, subsidence and collapses (steps) in the most superficial soils were exacerbated. This had the effect of the uninhabitability of thousands of homes. Since then, the city government undertook one of the largest reconstruction campaigns in the modern history of the city.

The toolbox for geotechnical engineers, intend to train them to detect aspects about differential subsidence, sudden deformations, and sinkholes. In the immediate aftermath of a major earthquake, they must also learn to catalog an additional symptom of deterioration: cracks and fissures in natural masses. This is particularly important as damage to build units can develop days after the seismic event if the openings in the soils and rocks are not properly treated (**Figure 10**). The registered details are stored and sent to the control center where are analyzed with a CART (classification tree) to determine if the site is on the "zero set" (cases where their conditions are on the top of risk levels and demands immediate actions).

The machine learning analysis is based in layers of Geo-descriptors that permits to qualify the susceptibility to sink-fracture in specific Mexico City regions. With the analysis of 6 variables (Geo-position, Soil heterogeneity, Groundwater Level, Urban loads, Type of foundations, Use of nearby streets) a CART (**Figure 11**) determines the relative influence of each one on the cracks' development and relates present conditions to a risk level. The user of the tool can request from the control center the result of the evaluation with CART so that he, when faced with any doubt about the state of the soils, can make a decision and qualify the situation.

Figure 10.
The vision of the user in the field is shown, a) in the Tláhuac area, in a serious crack, he observes the information on geotechnical zoning, b) check, in one of the most damaged areas, the repair of the drainage ducts, observes the optimal filling conditions according to the regional sinking map c) at site, where a leak and step deformation are present, infers which is the damaged section and record it in the application to call a specialized team.

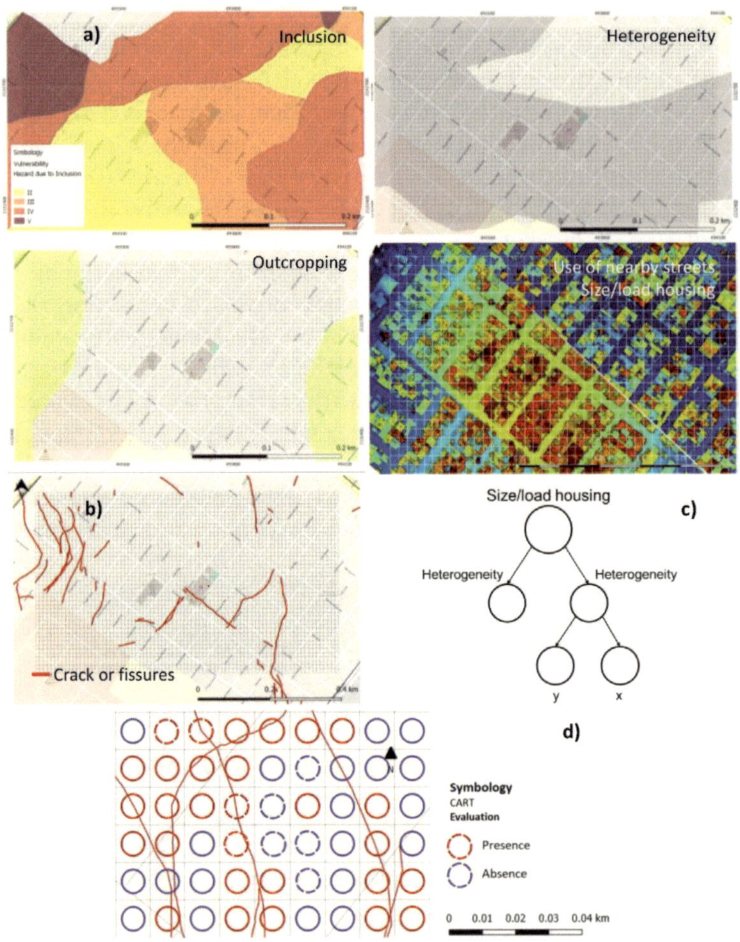

Figure 11.
CART for conclude about the susceptibility to crack. The model uses (a) information from maps of geo and anthropic properties and (b) survey of field damage. This tool (c) permits the user to qualify the susceptibility of a site to the development of a crack. (d).

For example, let us examine a site in the southern poor, and very susceptible to cracking, region in the metropolis. In **Figure 12** the user's vision when entering the site is shown. The characteristics of the breaks that must be recorded are geometry, materials (under consideration of soil evaluation according to the SUCS unified soil classification system), relative movement between the flanks of the crack/ step/ subsidence, among others. When the user asks for the CART response for this site, the geotechnical information (boreholes) that the AR-tool finds near the site is first shown and then the evaluation is presented with a disaggregated description of the factors that lead to that level of susceptibility. In the case exposed, the presence of a non-continuous thin layer of semi-rigid material embedded in the plastic clay, its proximity to the battery of water pumping wells (operating at relatively shallow depths) and the shape and depth of the rigid base, are the conditions that drives to the site to a high susceptibility to crack.

4.3 Recovery post-emergency

When the conditions of habitability and urban services have begun the path towards normalization, government administrators and practitioners (particularly engineers dedicated to the generation of infrastructure) should consolidate the

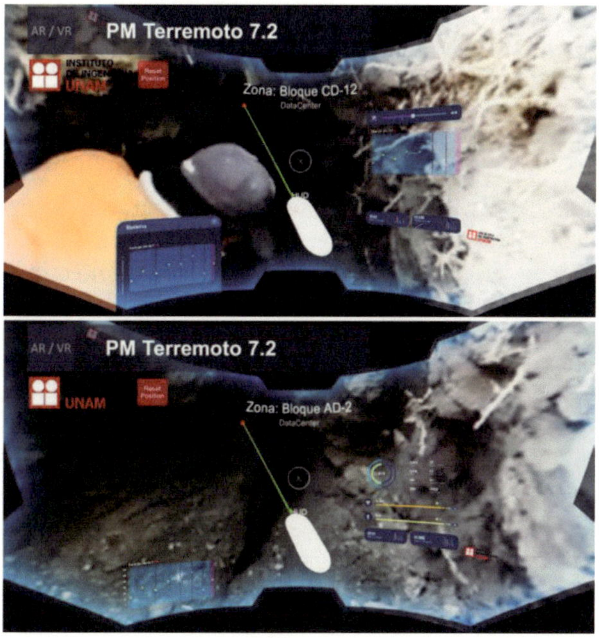

Figure 12.
Conditions at the entrance of a trench found under a foundation (3 floor house). The case is studied with the AR tool because of a call after finding a deep crack. The user of AR can measure the geometry and send/store this data. It also has information from nearby geotechnical surveys that help him interpret what observes. The videos of what happens are recorded for later analysis.

assistance programs (repair of minor damages in houses and buildings as well as attention to water pipelines) and larger-scale plans for the correction of structures that are considered not to follow what is established in the code or that require the application of complex engineering solutions (structural reinforcement, total or partial demolitions, complex restoration, control of settlements (fill), repair of sinkholes, etc.).

In this case the teams (engineers, designers, government, and urban planners) are provided with integral tools to evaluate causes-effects and directly determine how many resources the city and community needs to correct the risk situations and how to improve its resilience for future, which is one of the most effective long-term strategies. These virtual collaborative immersive spaces allow different experts from all over the instances to work together in the same virtual environment, creating enhanced and coordinated solutions. Resilience, poverty, security, among others, are social aspects that are presented in maps in superposition on the kind of damages and necessities of reparation. In this way the decisions about resources can be developed on a solidarity base that prioritizes the needs of the most marginalized communities. An additional aspect, very important for the NDM, is that one of the layers shown is the prediction of accelerations in the different geotechnical zones of the city when certain earthquakes attack the metropolis. With these interpretations and the effects, routes can be drawn for the adaptation of services and infrastructure to anticipate the potential scenarios of shutdown of functions in the event of a mega earthquake.

5. Conclusions

For gathering a more organic, equal, and inclusive world with no one left behind, there is an urgent need to transform the current unsustainable interactions within social ecological systems. The role of emerging technologies in achieving harmonious interactions is crucial. The tools presented in this investigation show how they can be used to enhance resilience to environmental disruptions, particularly earthquakes threat. It has been showed how Artificial Intelligence, VR and AR can improve natural disaster management closing the gap between knowledge and action.

VR-AR technology provides visual simulations to create a vivid first-person experience. Temporal, spatial, and social differences are lowered by immersing people into a certain location or experience. This immersion can be exploited in the three phases of NDM: preparedness, response, and recovery. Artificial Intelligence + VR-AR is an innovative addition to NDM as it provides a non-destructive and safe technique to recreate natural disasters. The simulation of future conditions using the parametric relationships found by NN improves the understanding of what kind of damage is caused, how to better prepare, and shows the possible ways to recovery.

NDM undoubtedly benefits from the adoption of VR-AR technologies. These tools are most advantageous and attractive when data, models and strategies are included from all possible sources involved in preparing for, responding to, and recovering from the effects of a natural phenomenon. The integration of geological, seismic, geotechnical, urban, service, social data, to name a few, in an interchangeable and visually stimulating format, directly impacts the effectiveness, cost and execution time of activities related to each stage of a NDM project. The proposal presented in this document was used as a prototype by some teams during the 2017 earthquake and showed its great potential to create routes that increased productivity and promoted more agile and socially supportive decision-making processes, reducing duplication of work and some traditional activities prone to big mistakes.

The requirement derived from the collaboration between actors from different branches of knowledge (from scientists, technologists, engineers, administrators, politicians to representatives of civil society) when an earthquake occurs in a large metropolis, and especially if it is of enormous complexity such as Mexico City, will increasingly force the use of tools that allow raising quality in i) training in predicted scenarios on responses of soils and rocks interacting with buildings and buried infrastructure, ii) the conceptualization of risk management and the monitoring of activities that affect the resilience of the most vulnerable populations and iii) the interpretations of behaviors, relationships, patterns and models with intelligent tools that allow exploring a multiparametric, highly dimensional, extremely complex natural/anthropic universe.

The aspect of absolute and partial immersion, the congruent and integral intelligent modeling, the exploration of the programming and the risk management tools, as well as the interoperability, interactions and exchanges between the actors that generate huge amounts of data in all types and formats, without a doubt, it is a monumental challenge not only to the exercise of those who are in charge of NDM but to the adjustments to the analytical and mental connections, with which we have historically directed the responses to the demands of emergency situations, however the recent experience has shown that the application is effective and promising.

Acknowledgements

The authors would like to express their very great appreciation to the Government of Mexico City and to the Commission for the Reconstruction. Also, we would like to thank the following company for its assistance with the collection of the data and to share experiences: LOHRE S.A. de C.V.

Conflict of interest

The authors declare no conflict of interest.

Author details

Silvia García[1*], Paulina Trejo[1] and Alberto García[2]

1 Engineering Institute UNAM, CDMX, México

2 GEOIntelligent Systems Laboratory, CDMX, México

*Address all correspondence to: sgab@pumas.iingen.unam.mx

References

[1] Deng, Z., Yu, Y., Yuan, X., Wan, N. and Yang, L., 2013. Situation and development tendency of indoor positioning. China Communications, 10(3), pp.42-55.

[2] Li, Y., 2016. Integration of MEMS sensors, WiFi, and magnetic features for indoor pedestrian navigation with consumer portable devices (Doctoral dissertation, University of Calgary).

[3] Anthes, C., R. J. García-Hernández, M. Wiedemann and D. Kranzlmüller (2016). "State of the art of virtual reality technology." In: Proceedings of the IEEE Aerospace Conference, Big Sky, MT, March 2016.

[4] Freina, L. and Ott, M., 2015, April. A literature review on immersive virtual reality in education: state of the art and perspectives. In The International Scientific Conference eLearning and Software for Education (Vol. 1, p. 133). "Carol I" National Defence University.

[5] Cipresso P., Irene Alice Chicchi Giglioli, Mariano Alcañiz Raya and Giuseppe Riva. (2018). The Past, Present, and Future of Virtual and Augmented Reality Research: A Network and Cluster Analysis of the Literature. Front. Psychol., https://doi.org/10.3389/fpsyg.2018.02086

[6] Joda Tim, Gallucci German and Wismeijer Daniel. (2019). Augmented and virtual reality in dental medicine: A systematic review. Computers in Biology and Medicine 108:93-100. DOI: 10.1016/j.compbiomed.2019.03.012

[7] Hsu, E.B., Li, Y., Bayram, J.D., Levinson, D., Yang, S. and Monahan, C., 2013. State of virtual reality-based disaster preparedness and response training. PLoS currents, 5.

[8] Nunes, I.L., Lucas, R., Simões-Marques, M. and Correia, N., 2018, August. An augmented reality application to support deployed emergency teams. In Congress of the International Ergonomics Association (pp. 195-204). Springer, Cham.

[9] S.L. Handy, M.G. Boarnet, R. Ewing, R.E. Killingsworth, How the built environment affects physical activity: Views from urban planning, Am. J. Prev. Med. 23 (2002) 64– 73. https://doi.org/10.1016/S0749-3797(02)00475-0.

[10] J. Wilson, A. Oyola-Yemaiel, The evolution of emergency management and the advancement towards a profession in the United States and Florida, Saf. Sci. 39 (2001) 117-131. https://doi.org/10.1016/S0925-7535(01)00031-5.

[11] F.H. Norris, S.P. Stevens, B. Pfefferbaum, K.F. Wyche, R.L. Pfefferbaum, Community resilience as a metaphor, theory, set of capacities, and strategy for disaster readiness, Am. J. Community Psychol. 41 (2008) 127-150. https://doi.org/10.1007/s10464-007-9156-6.

[12] B.L. Murphy, Locating social capital in resilient community-level emergency management, Nat. Hazards. 41 (2007) 297-315. https://doi.org/10.1007/s11069-006-9037-6.

[13] Gaillard J, Mercer J. From knowledge to action: Bridging gaps in disaster risk reduction. Prog Hum Geogr. 2013;37(1):93-114. doi:10.1177/0309132512446717

[14] de Regt A, Barnes SJ, Plangger K. The virtual reality value chain. Bus Horiz. 2020;63(6):737-748. doi:10.1016/j.bushor.2020.08.002

[15] Virtual Reality Market by Segment (Consumer, Enterprise, Industrial, Government), Equipment (Hardware, Software, Components), Applications, and Solutions 2019 – 2024.; 2019.

[16] Hartmann T. Interview IEPC Consultancy (2020).

[17] Velev D. Interview IEPC Consultancy (2020).

[18] García S. Interview IEPC Consultancy (2020).

[19] Belleman R. Interview IEPC Consultancy (2020).

[20] Nijland T. Interview IEPC Consultancy (2020).

[21] Fauville G. Interview IEPC Consultancy (2020).

[22] Perriguey G. Interview IEPC Consultancy (2020).

[23] van Wijngaarden M. Interview IEPC Consultancy (2020).

[24] Deepak M. Interview IEPC Consultancy (2020).

[25] Sermet Y. Interview IEPC Consultancy (2020).

[26] Calil J. Interview IEPC Consultancy (2020).

[27] Boersma K. Interview IEPC Consultancy (2020).

[28] Ortiz H, Herrera Terrada V. Interview IEPC Consultancy (2020).

[29] Schwarz, A., Binetti, J.C., Broll, W. and Mitschele-Thiel, A., 2016. New technologies and applications in international crisis communication and disaster management. The handbook of international crisis communication research, 43, p.465.

[30] Jung, S.U., Cho, H. and Jee, H.K., 2016, November. An AR-based safety training assistant in disaster for children. In SIGGRAPH ASIA 2016 Posters (p. 49). ACM.,

[31] Vichitvejpaisal Pongsagon, Nutawut Yamee and Panukom Marsertsri. (2016). Firefighting simulation on virtual reality platform. Conference: 2016 13th International Joint Conference on Computer Science and Software Engineering (JCSSE). DOI:10.1109/JCSSE.2016.7748912

[32] Alharthi Rajwa Abdullah, Benjamin Guthier, Abdulmotaleb El Saddik. (2018). Recognizing Human Needs During Critical Events Using Machine Learning Powered Psychology-Based Framework. IEEE Access PP (99):1-1. DOI:10.1109/ACCESS.2018.2874032

[33] Sharma, S., Devreaux, P., Sree, S., Scribner, D., Grynovicki, J. and Grazaitis, P., 2019. Artificial intelligence agents for crowd simulation in an immersive environment for emergency response. Electronic Imaging, 2019(2), pp.176-171.

[34] S. García, P. Trejo and A. García. (2021). Intelligent Virtual Reality of a Metropolis Devastated by Earthquakes, 17th World Conference on Earthquake Engineering, 17WCEE.

[35] Itamiya, T., Tohara, H. and Nasuda, Y., 2019, March. Augmented Reality Floods and Smoke Smartphone App Disaster Scope utilizing Real-time Occlusion. In 2019 IEEE Conference on Virtual Reality and 3D User Interfaces (VR)(pp. 1397-1397). IEEE.

[36] Tsai, M.K. and Yau, N.J., 2013. Improving information access for emergency response in disasters. Natural Hazards, 66(2), pp.343-354.

[37] Veas, E., Grasset, R., Ferencik, I., Grünewald, T. and Schmalstieg, D., 2013. Mobile augmented reality for environmental monitoring. Personal and ubiquitous computing, 17(7), pp.1515-1531.

[38] Luchetti, G., Mancini, A., Sturari, M., Frontoni, E. and Zingaretti, P., 2017.

Whistland: An augmented reality crowd-mapping system for civil protection and emergency management. ISPRS International Journal of Geo-Information, 6(2), p.41.

[39] Mirauda, D., Erra, U., Agatiello, R. and Cerverizzo, M., 2018. Mobile augmented reality for flood events management. Water Studies, p.47.

[40] Fedorov, R., Frajberg, D. and Fraternali, P., 2016, June. A framework for outdoor mobile augmented reality and its application to mountain peak detection. In International Conference on Augmented Reality, Virtual Reality and Computer Graphics (pp. 281-301). Springer, Cham.

[41] Reyes, M.E.P. and Chen, S.C., 2017, April. A 3D virtual environment for storm surge flooding animation. In 2017 IEEE third international conference on multimedia big data (BigMM)(pp. 244-245). IEEE.

[42] Ready, M., Dwyer, T. and Haga, J.H., 2018. Immersive Visualisation of Big Data for River Disaster Management. In AIST.

[43] Haynes, P., Hehl-Lange, S. and Lange, E., 2018. Mobile Augmented Reality for Flood Visualisation. Environmental modelling & software, 109, pp.380-389.

[44] Macchione, F., Costabile, P., Costanzo, C. and De Santis, R., 2019. Moving to 3-D flood hazard maps for enhancing risk communication. Environmental modelling & software, 111, pp.510-522.

[45] P. Milgram, F. Kishino, A taxonomy of mixed reality visual-displays, in: 1st Int. Work. Networked Real. (NR 94), Institute of Electrical and Electronics Engineers Inc., 1994: pp. 1321-1329. https://doi.org/10.1109/ISPAN.1997.645049

[46] X. Li, W. Yi, H.L. Chi, X. Wang, A.P.C. Chan, A critical review of virtual and augmented reality (VR/AR) applications in construction safety, Autom. Constr. 86 (2018) 150-162. https://doi.org/10.1016/j.autcon.2017.11.003.

[47] R.T. Azuma, A survey of augmented reality, Presence Teleoperators Virtual Environ. 6 (1997) 355-385. https://doi.org/10.1162/pres.1997.6.4.355

[48] J. Carmigniani, B. Furht, M. Anisetti, P. Ceravolo, E. Damiani, M. Ivkovic, Augmented reality technologies, systems and applications, Multimed. Tools Appl. 51 (2011) 341-377. https://doi.org/10.1007/s11042-010-0660-6

[49] Isdale, J. (1993). What is virtual reality? A homebrew introduction. Retrieved November 12, 2005. [Online]. Available: http://whatis.techtarget.com

[50] Haag, S.; Cummings, M., & Dawkins, J. (1998). Management Information Systems for the Information Age. Irwin/McGraw Hill, ISBN 0-07-025465-6, New York.

[51] Kim, J., Park, S., Yuk, K., Lee, H. and Lee, H. (2000). Virtual reality simulations in physics education. Interactive Multimedia Electronic Journal of Computer-Enhanced Learning [Online]. Available: http://imej.wfu.edu/articles/2001/2/02/index.asp

[52] Williams, B.K., & Sawyer, S.C. (2001). Using Information Technology: A Practical Introduction to Computers and Communications. McGraw Hill, ISBN 0-07-239803-5, New York.

[53] Onyesolu, M.O. (2009). Virtual reality laboratories: The pedagogical effectiveness and use in obtaining cheap laboratories using the computer laboratory, Journal of Science

Engineering and Technology, Vol. 16, No.1, (March 2009) 8679-8689, ISSN 1117-4196.

[54] Onyesolu, M.O. & Akpado, K.A. (2009). Virtual reality simulations in computer engineering education. International Journal of Electrical and Telecommunication Systems Research, Vol. 3, No.3, (July 2009) 56-61, ISSN 0795-2260.

[55] Isdale, J. (1998). What is virtual reality? A web-based introduction. Retrieved November 12, 2005. [Online]. Available: http://whatis.techtarget.com/definition//0.sid9_gci213303,00.html

[56] Aukstakalnis, S. & Blatner, D. (1992). Silicon mirage: The art and science of virtual reality. Peachpit Press, Berkley.

[57] Cruz-Niera, C. (1993). Virtual reality overview. Proceeding of ACM SISGRAPH 93 Conference on Applied Virtual Reality, Anaheim, California.

[58] Lane, C. (1993). Display Technologies. Encyclopedia of Virtual Environments. World Wide Web URL: http://www.hitl.washington.edu/scivw/EVE/I.A.1.Displays.html

[59] P. Milgram, and A. F. Kishino, Taxonomy of Mixed Reality Visual Displays IEICE Transactions on Information and Systems, E77-D (12), pp. 1321-1329, 1994.

Chapter 7

Advances in Spatially Faithful (3D) Telepresence

Seppo Valli, Mika Hakkarainen and Pekka Siltanen

Abstract

Benefits of AR technologies have been well proven in collaborative industrial applications, for example in remote maintenance and consultancy. Benefits may also be high in telepresence applications, where virtual and mixed reality (nowadays often referred as extended reality, XR) technologies are used for sharing information or objects over network. Since the 90's, technical enablers for advanced telepresence solutions have developed considerably. At the same time, the importance of remote technologies has grown immensely due to general disruption of work, demands for reducing travelling and CO2, and the need for preventing pandemics. An advanced 3D telepresence solution benefits from using XR technologies. Particularly interesting are solutions based on HMD or glasses type of near-eye-displays (NED). However, as AR/VR glasses supporting natural occlusions and accommodation are still missing from the market, a good alternative is to use screen displays in new ways, better supporting e.g. virtual meeting geometries and other important cues for 3D perception. In this article, researchers Seppo Valli, Mika Hakkarainen, and Pekka Siltanen from VTT Technical Research Centre of Finland describe the status, challenges, and opportunities in both glasses and screen based 3D telepresence. The writers also specify an affordable screen based solution with improved immersiveness, naturalness, and efficiency, enhanced by applying XR technologies.

Keywords: 3D telepresence, spatial faithfulness, remote interaction, XR technologies, AR/VR glasses

1. Introduction

This article compiles lessons learned by the writers from more than a decade of telepresence related research. The article is a review by nature, but due to the number of the described topics, the presentation is not tutorial in all parts, but relies on prior knowledge by its readers, and/or their interest to learn more e.g. from the given references. Based on the reviewed status and enablers, the writers reason and define a practical and affordable 3D telepresence solution based on screen displays. In this solution, efficient 3D capture and low bitrate streaming is an important enabler both for communication and XR functionalities. Essentially the same technical solutions can also be used for remote support applications in industry, e.g. for 3D monitoring, maintenance, control, analysis, and augmentation.

The outline of the paper is as follows. In Chapter 2, we introduce the 3D telepresence topic, describe main factors of spatial faithfulness, and give few examples of existing approaches. Several of the references are to our own patent publications, which have not been published as papers. Chapter focus is in the requirements and challenges of supporting 3D geometries and perception, as perceived in real-world encounters (face-to-face).

In the future, glasses type of displays will likely be the best to support immersion, mobility and freedom of viewpoint. However, still today, glasses are still lacking many important properties, and have many defects limiting perceived quality, time of use, and user acceptance. At the same time, using screen displays is the most common way of supporting visual interaction in teleconferencing solutions. Correspondingly, we wanted to find out whether a simple screen-based telepresence solution could support 3D perception and XR functionalities with improved naturalness, quality, and usability. The answer seems to be positive, and in Chapter 3, we give a draft specification for such a system.

An important enabler both for communication and AR functionalities is efficient 3D capture and streaming. Further, in Chapter 3, an implementation applying existing coding methods is described together with some simulation results. Despite our demarcation to screen based solutions, we also discuss the possibilities and future of glasses based approaches. In Chapter 4, we describe ways of enhancing 3D perception and XR functionalities of the basic solution. Future improvements may include also supporting natural eye-focus by accommodative displays. Finally, Chapter 5 summarizes our findings.

2. Basics and examples of spatially faithful telepresence solutions

2.1 Spatial faithfulness supports naturalness in perception

3D telepresence solutions aim to support natural perception in 3D – sc. spatial faithfulness – better than video conferencing systems [1]. Basic problem in videoconferencing systems is the lack of support for eye contact [2]. In flat screen based solutions, it stems for example from the displacement or offset between a display (showing a counterparty's face) and camera (counterparty's eyes) of a videoconferencing terminal. Note, that although eye contact is one of the early goals for 3D telepresence solutions, it is still not supported in most of the existing telepresence systems.

Hydra system (**Figure 1**) is an early approach for supporting spatial faithfulness in telepresence [3, 4]. With a mesh of connections and a separate (proxy) terminal for each remote counterpart, it aims to support virtual lines-of-sight between participants. With small displays and small camera-display offset(s), participants are

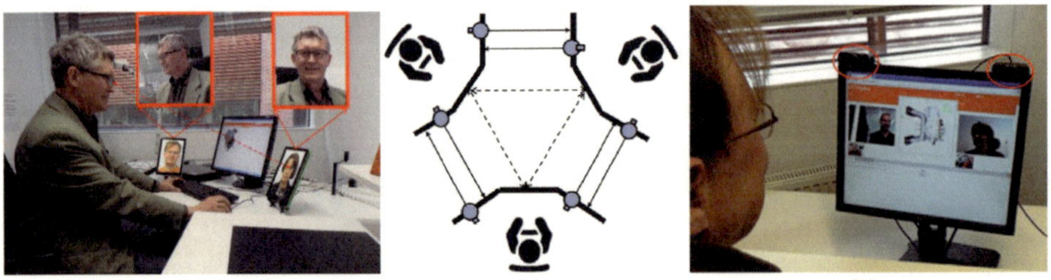

(a) (b) (c)

Figure 1.
a) VTT idea for a Hydra system using tablets for a communication space and a computer display for a collaboration space, b) corresponding connections between cameras and displays, c) practical implementation showing all contents on one display (cameras on top corners are indicated by red circles) [3].

able to get an approximate eye contact with each remote participant, and in certain conditions, even have a shared understanding of each other's relative positions.

Note that perceiving eye contact does not mean full gaze awareness, i.e. understanding of also other, intermediate eye directions. In a traditional Hydra system, terminals for each remote party can be placed independently by each local participant, which easily results with inconsistent meeting geometries across meeting sites, and thus also inconsistent positions and eye directions of the parties.

Note that in principle, any position of displays can support virtual lines-of-sight between participants. The situation can be compared to private house residents seeing their neighbors through windows, which, however, is the more unlikely the more of neighbors like to communicate with each other.

The easiest solution unifying a virtual meeting geometry between participants is to position proxy terminals in the same relative order into vertices of an equilateral polygon (e.g. a triangle, square, pentagon, hexagon, etc.). This naturally restricts the seating of participants more than in a face-to-face meeting.

A more recent screen based solution is Viewport by Zhang et.al [5]. In the Viewport system, high-quality 3D models are formed for each user in real time, and extracted and embedded into a common virtual geometry. Using 3D models enables correcting camera-display offset, and supporting depth perception by stereoscopy. The system supports eye contact between three sites, with one user at each site. In particular, limiting the number of sites into only few is a factor limiting the usability of corresponding solutions.

Natural perception of depth and distances belongs to the factors of spatial faithfulness. Note that this is not possible using 2D displays lacking depth, and strictly speaking not even with stereoscopic 3D (S3D) displays, whether multiplexed, polarized, or autostereoscopic, due to their incapability to support natural focus/accommodation (suffering from the sc. vergence-accommodation conflict, VAC [6]).

2.1.1 Advances by XR technologies and computer games

Note that spatial faithfulness is an inherent requirement in XR visualization, which aims at replacing, in a seamless way, parts of a physical view with virtual elements (or vice versa). XR visualization can as well be used for rendering models (cf. avatars) or visual reconstructions of human participants into a participant's view. Correspondingly, for more than two decades, developing enablers for XR has also advanced 3D telepresence solutions. These enablers include e.g. sensors for 3D capture, coding and streaming methods, low latency networks, tracking and detection for XR, camera and user positioning, motion capture and tracking methods, new display technologies, and general advances in algorithms and processing power.

In the same way, developing game technologies has advanced 3D telepresence solutions based on virtual modeling and rendering, here denoted as Virtual World (VW) approaches. Traditionally, in VW approaches, visual content has been modeled/produced in advance, and rendering of the content is based on real-time transfer of parameters for viewpoint and object positions, dynamic 3D shapes and poses (animation), etc. Although VW approaches are rather common and have their specific benefits, their description is omitted in our presentation, focusing on rendering of photorealistic real-time captures. This focus is reasoned in more detail in Chapter 2.5.

A recent example of a photorealistic telepresence solution based on advanced 3D displays is Google Project Starline (https://blog.google/technology/research/project-starline/). A good example of a 3D telepresence system based on AR/VR (XR) visualization is MS Holoportation [7] (cf. https://www.youtube.com/watch?v=7d59O6cfaM0). Both of them are quite impressive but obviously also complicated and costly. In this article, we aim to define a more economical solution with good sides of both photorealism and XR visualization. Note that even a lone talking head on a screen - whether camera captured or 3D modeled - may well be a value-adding functionality. Communication and attractiveness may namely be supported by using e.g. a speech-controlled, look alike or anonymous virtual head (cf. https://remoteface.ai/ and a video at: https://www.youtube.com/watch?v=prpPqwV5Weo).

2.2 Remarks on mobility and serving with viewpoints

In above, Hydra system was described as an early attempt towards spatially faithful 3D telepresence. Using such full-mesh approach, and by making restricting assumptions on participant positions ("seating order"), all participants may perceive eye directions and participant positions consistently, however, apart from solving the disturbing camera-display offset. Furthermore, a regular setup with fixed camera and display positions naturally limits the mobility of participants within their meeting sites.

Note, that although a participant position is fixed, a whole meeting room with its occupant may move virtually. In [8], a solution is described for compiling captures of regular sensor and display setups into a landscape, enabling participants to mingle together with their meeting spaces within each other, like people in a cocktail party (**Figure 2**). For example, a capture setup in a square or hexagonal formation can be used. Writers of this paper are however not aware if someone has implemented and tested such arrangement.

Let us consider a Hydra setup a little bit further. If a Hydra system with all its terminals was in one large hall or open space, a participant is able to switch (walk) between different sites and perceive spatial faithfulness in each of them separately, i.e. participant mobility is supported in discrete locations.

User mobility may be supported in principle at any viewpoint when using near-eye glasses (NEDs) for viewing. Limitations set by fixed camera positions can be relieved using a setup of multiple 3D sensors, or multiple cameras in an array. For the latter, solutions based on wall-mounted camera arrays or moving cameras are described in [9, 10], correspondingly. Arbitrary viewpoints can be supported to remote spaces, provided that complete and real-time enough 3D reconstructions of those spaces are available, or that one of the multiple cameras provides the viewpoint along a desired line-of-sight (**Figure 3**).

For serving viewpoints from varying positions, i.e. receiving viewpoints on-demand, the system needs to deliver participant positions between sites in unified coordinates (a unified geometry). Renderings of remote participants need to be

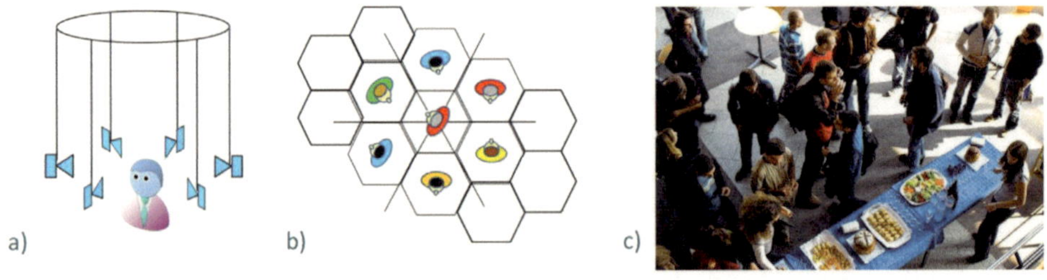

Figure 2.
Fixed capture setups can support collaboration in dynamic 2D landscapes: a) capture setup in hexagonal grid, b) captures arranged into a tessellated landscape, enabling c) moving with user spaces like people in a cocktail party [8] (image c) is creative commons image by Lucas Maystre from Renens, Switzerland - 053/365: Apéro au forum).

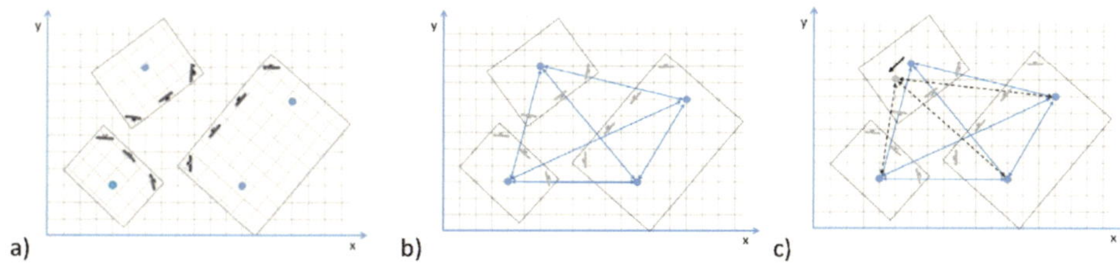

Figure 3.
Idea of bringing real-world meeting sites into a common geometry: a) three sites with users (dots) captured by RGB-D cameras in local coordinates, b) lines-of-sight between users in a unified coordinate system, and c) supporting lines-of-sights (new viewpoints) for a moving participant.

compiled in each participant view correspondingly. A viewer's head orientation needs to be detected and tracked to serve him/her with a correct part (frustum) of the compiled scene.

The writers have disclosed several inventions using the above described approach [8–10]. Note that instead of delivering visual information as large and bitrate consuming 3D volumes, a visual stream may be a video, a stereoscopic video, or a video-plus-depth stream (V + D, [11]) allowing forming a stereoscopic video. When using the viewpoint-on-demand approach in this way, the bitrate requirement may be much lower than when streaming 3D volume data. Note that Chapter 2.4 will describe the viewpoint-on-demand approach in more detail.

As a summary, spatially faithful telepresence solutions aim to support real-world-like geometries among participants. However, as will be explained in Chapter 3.2, both virtual and photorealistic approaches may be feasible even without such support, and even much easier. For example, perception of depth and motion parallax can be supported without forming and maintaining perfectly unified virtual meeting geometries.

2.3 Transmitting and displaying volume videos (3D streams)

Ideally, 3D reconstructions are coded, delivered and displayed as real-time 3D streams. However, this is very challenging e.g. due to high computation power, and very high bitrate it requires. The benefits of volume videos include more freedom in choosing one' viewpoint (cf. motion parallax and alternative viewpoints) and support for multiple local viewers. However, both capturing participant spaces and supporting viewing with glasses bring considerable complexity to this approach.

Viewing from various viewpoints may be supported also using multi-view streaming and display methods. However, without capture and delivery of user positions (cf. knowledge of a mutual meeting geometry), users need to choose their position accurately among a priori specified locations e.g. in order to perceive correct eye contact(s). Several approaches are using this approach, although simplified by reducing the volume being supported. Multi-view video coding methods and standards are available and applicable for this [12], but more advanced (real-time) 3D coding method are still under development e.g. by MPEG. Special 3D displays are already available, supporting different 3D viewpoints for multiple local viewers.

In the future, by advances in transmission and display (e.g. using light fields), real-time streaming and display of 3D volumes becomes more feasible. An apparent benefit of these solutions is that simultaneous viewers can see the 3D content from their individual viewpoints, like in the real world.

2.4 Viewpoint-on-demand – simplifying spatially faithful solutions in low latency network

5G seems to provide enough bitrate with low latency for future 3D telepresence services. According to Ronan McLaughlin (Ericsson, Ltd.), the 5G system design parameters specify a system capable of delivering an enhanced mobile broadband (eMBB) experience, in which users should experience a minimum of 50–100 Mbps everywhere, and see peak speeds greater than 10 Gbps, with a service latency of less than 1 ms, while moving at more than 300 miles/h! (https://broadbandlibrary.com/?s=5G+Low+Latency+Requirements).

Spatial faithfulness requires a shared geometry between meeting participants. In order to form and maintain such geometry, user positions need to be detected, tracked, and delivered at each moment. In addition, defined by the geometry, 3D data from several remote sites needs to be streamed to each local viewer, and compiled in a unified 3D representation. If each of the 3D captures is a full reconstruction of the corresponding 3D space, the overall bitrate requirement for rendering each view becomes huge. This may be even too much for the emerging 5G network (or at least costly). In addition to high bitrate, a very important potential of 5G network is its low latency (cf. the above figures by Ronan McLaughlin).

Most of the existing approaches for 3D telepresence are aiming to capture, encode and stream visual data of at least partial 3D volumes, which then can be seen from various viewpoints in the receiver. However, a person is able to see only from one (binocular) viewpoint at a time, which means that at each moment, there is need to see only one projection to a 3D volume. Assuming that a viewer's motions are moderate, and that a low latency network like 5G is available for data streaming, the complexity of a 3D telepresence system can be considerably reduced by streaming only video-plus-depth (V + D) projections from tracked viewpoints. Valli and Siltanen have made several telepresence inventions using this sc. viewpoint-on-demand (VoD) approach [8–10]. In particular, applying augmented reality to 3D telepresence is described in inventions [13, 14]. Note that an example of our recent 3D streaming implementation is given later in Chapter 3.4, and using the solution for supporting XR functionalities is described in more detail in Chapter 4.

Note that synthesizing viewpoints e.g. for supporting motion parallax and correcting camera offset for eye contact may be possible without ordering new data and experiencing a corresponding two-way delay as a result. An obvious way of reducing the need for delivering data for new viewpoint is to use multiple-viewpoint video coding instead of video-plus-depth (V + D) data [15]. This allows more freedom for viewpoint changes within the received stream. For examples of reducing viewpoint orders, see also several inventions by Valli and Siltanen on synthesizing stereoscopic or accommodative (MFP) content for small viewpoint changes [16, 17].

2.5 Photorealistic vs. virtual world (VW) approaches

Note that serving with arbitrary viewpoints may be easier in VW approaches, where virtual camera views are formed to a shared virtual meeting space (VW), using the knowledge of each viewer's pose (tracked by VR glasses, or defined by a participant e.g. by a mouse). However, virtual environments with animated avatars are less natural, and may even alienate a participant by causing the sc. uncanny valley effect. On the other hand, using modeled avatars for participants provides the possibility for their anonymity or role-play, which is an obvious benefit in some use cases and services.

Using a virtual world approach is a viable option used by several service vendors (see e.g. references in https://en.wikipedia.org/wiki/Virtual_world). In VW approaches, meeting spaces are typically modeled in advance, and as much as possible, also delivered to the participants in advance. Coding and delivering corresponding 3D information may be based e.g. on hierarchical volume coding methods like OctoMap by Hornung et al. [18]. Despite partly in-advance delivery for the meeting space, a lot of accurate motion and animation parameters remain to be delivered, and graphical processing to be made for local renderings (e.g. for forming viewing frustums). As a result, although possibly lighter than photorealistic approaches, VW approaches are by far not simple either.

In a photorealistic approach, capturing, forming, and delivering reconstructions of 3D volumes and human participants is more challenging, although the meeting spaces can likely be modeled in advance. Once formed, 3D reconstructions can be viewed like in VW approaches, using NEDs or HMDs. Naturally, hybrid solutions combining photorealistic and VW approaches are also possible. For example, 3D modeled environments may be used instead of captured meeting spaces, and XR functionalities can be used for augmenting avatars.

Figure 4 (cropped screenshots of YouTube videos by the courtesy of Oliver Kerylos) gives examples of hybrid (XR) approaches, showing real-time captured participants in 3D modeled meeting spaces.

As seen in **Figure 4**, a particular challenge in this approach is that a glasses display covers a person's face, which prevents a viewer from seeing his face and perceiving eye contact. Solutions for this have however since been described in literature, based on real-time manipulation of facial areas [19].

As a short summary, main approaches for 3D telepresence can be classified into the following four classes (**Table 1**).

Note that the quadrants of the table correspond to the classical reality-virtuality continuum by Milgram and Kishino [20]. Current videoconferencing and Virtual World approaches correspond to the real and virtual ends of this continuum, and hybrid approaches respectively to intermediate positions labeled as augmented reality (AR) and augmented virtuality (AV). Note that in parallel to the commonly used term "Mixed Reality" (MR), also the term "Hybrid Reality" (HR) was discussed in [20]. Recently the term "Extended Reality" (XR) has gained popularity much in the same meaning.

In an augmented reality (AR) approach, a virtual avatar is representing each remote participant in a local environment. This requires capturing a remote participant's facial and body gestures and animating the avatar correspondingly. Respectively, in an augmented virtuality (AV) approach, photorealistic 3D captures of participants are made and delivered in real-time to a virtual meeting space (VW).

Figure 4.
Examples of XR approaches: a) person capture in a virtual space (2014), b) interaction in virtual space (2016), c) XR collaboration in virtual space (2012). (see https://www.youtube.com/channel/UCj_UmpoD8Ph_EcyN_xEXrUQ).

	Real space	**Virtual space**
Real human	**Photorealistic 3D approaches:** • Real-time capture of participants and spaces (cf. videoconferencing) • Challenges in supporting: o Shared geometry and depth for natural perception o Participant/user mobility o Avoiding use of HMD/glasses	**Hybrid (AV) approaches:** • 3D participant captures in VW • Challenges: o Real-time 3D capture, reconstruction, and delivery of participants o HMD/glasses obstructing faces
	Mutual challenge/opportunity while showing people and spaces: **Supporting remote XR interactions**	
Virtual human (avatar)	**Hybrid (AR) approaches:** • Avatars in physical spaces Challenges: o Capturing participants for animating avatars with natural motions and facial features o HMD/glasses obstructing faces	**Virtual (VR) approaches:** • Scalable, spatially faithful virtual (VW) solutions exist • Challenges: o Avoiding unnaturalness o Supporting photorealism o HMD/glasses obstructing faces

Table 1.
Main approaches for 3D telepresence.

In our case, hybrid approaches are particularly interesting. Compared to local (traditional) XR visualizations, combining real and virtual elements over distances (i.e. remote XR) causes particular challenges. These are discussed in more detail later in Chapter 2.7 and 4.1.

2.6 About using glasses and screen displays

AR/VR glasses or head-mounted displays (HMD) – together referred to as near-eye displays (NED) – can in principle support full (sc. 6DoF) freedom in choosing ones viewpoint to the displayed 3D content. Naturally, this requires also enough physical space, and precise tracking of user's motion and orientation (sc. pose).

HMDs (VR glasses) are well accepted for playing immersive and interactive computer games, and are commonly used when using VW-based telepresence and online platforms. However, they are still challenged by resolution, weight, and lack of support for natural focus (accommodation), causing discomfort and nausea when viewing stereoscopic content [6, 21, 22].

Optical see-through (OST) AR glasses (cf. MS HoloLens) have succeeded best in XR applications, but in addition to sharing the above challenges of HMDs, they are lacking natural occlusions, e.g. the ability to block a real background by augmentations, which makes AR objects to appear translucent. When augmenting natural views with 3D objects or other visual elements, closer objects should in general occlude those further away. Real and virtual objects may be in any order, meaning that foreground objects need to occlude the background whether they are real-world captures or virtual renderings [23]. These sc. mutual occlusions are especially difficult to support by optical-see-through (OST) glasses - like MS HoloLens.

In telepresence use, a serious drawback of both HMDs and OST AR glasses is that they block their user's face, which makes it difficult to see a participant's facial features and eye-directions, either when animating an avatar in virtual approaches, or when viewing photorealistic captures. Correspondingly, although advanced considerably in recent years, NEDs are not yet good enough to be generally accepted and applied to 3D telepresence.

On the other hand, screen displays have developed by size, accuracy, and economy. While usually seen from some distance, they are e.g. less prone for perceiving VAC in stereoscopic viewing. Naturally, they restrict choosing ones viewpoints, and in XR, make it about impossible to mix remote and local content naturally in depth dimension (except when mirroring a local environment with augmentations). In short, screen displays are less immersive, but more easy to view than NEDs.

While the size and accuracy of screen displays is growing, and the distance of viewing them is reducing, supporting accommodation may become necessary also with screen displays. However, existing external accommodative displays support viewing either from very fixed viewpoints, or suffer from other severe limitations in rendering (cf. lack of colors, brightness and occlusions when using e.g. holographic volume displays).

2.7 About XR and its role in 3D telepresence

Spatial faithfulness is an inherent requirement in XR visualization, which aims at replacing, in a seamless way, parts of a physical view with virtual elements. XR visualization can as well be used for rendering 3D models (cf. avatars) or visual reconstructions of human participants into a participant's view. Correspondingly, for more than two decades, developing enablers for XR has correspondingly advanced also 3D telepresence solutions. VTT has made a lot of research in these topics, and examples of results can be found e.g. at web (http://virtual.vtt.fi/virtual/proj2/multimedia/) and in YouTube (www.youtube.com/user/VTTAugmentedReality).

VTT made also an early implementation of MR telepresence (MR Conferencing) in 2008–2009 [24]. The implementation supported participation to telepresence sessions using normal videoconferencing terminals and screens, and a VR space (SecondLife) with avatars. Registration of avatars to a real space used visual markers, which at that time was the main approach in AR visualization. Note that today's MR telepresence implementations do not necessarily differ too much from the VTT example, except using feature based tracking instead of markers (**Figure 5**).

Traditionally, making 3D captures at the target location ("on the spot") has been the only way to *support precisely* either positioning or viewing AR content in the location. By precisely, we mean that AR content can be bound to both shapes and textures (i.e. to precise visual context). This has meant making 3D capture and reconstruction locally, as an offline and in-advance process. Correspondingly, remote production and positioning of XR objects - without knowledge of local visual context - has been based on: 1) assumed textures (e.g. on known/assumed markers/pictures/objects/color patterns), 2) locally scanned shapes when displaying (e.g. physical delimiters like floor and wall panes), or 3) actions by the viewer (e.g. positioning of avatars and talking heads for communication).

However, tele interaction and XR applications can be supported better if 3D capture for making augmentations is enabled also from remote and more in real-time. This is possible by using efficient 3D capture, coding and streaming methods. The importance of efficient and high quality 3D streaming and interaction is growing fast

Figure 5.
MR conferencing between virtual and real spaces: a) second life view (screenshot), b) real life (augmented video).

due to the transformation towards distributed industrial processes, and having at the same time needs for reducing physical travels. Writers of this paper have got successful results in applying standard coding methods into real-time streaming of video-plus-depth data from RGB-D sensors (including means of supporting high enough pixel dynamics for the depth sensor data). These are presented later in Chapter 3.4.

As a natural trend in 3D telepresence implementations, there is a need for increasing accuracy (pixel dynamics) and resolution in 3D reconstruction. Following the progress in industrial applications, Lidar sensors are likely to become also into use in telepresence solutions. In addition to now common point cloud coding and transmission (e.g. using octrees [18]), this will likely require new coding methods which – in addition or instead of point clouds – support efficiently real-time transmission and visualization of high-quality surfaces and color textures (cf. approaches used with RGB-D sensors).

2.8 Focus of the research and the rest of our paper

Most of existing telepresence solutions are either photorealistic or virtual, i.e. fall into the first and fourth quadrants in **Table 1**. Hybrid approaches mix real and virtual components (for either participants or spaces) meaning that they are XR approaches (cf. discussion in Chapter 2.5). Note that in telepresence, augmenting remote participants, spaces or objects occurs over network, meaning that it is about remote XR (cf. Chapter 2.7), which requires delivering more position and 3D data than in traditional AR, both for augmentation and viewing.

Further, although augmented content can be viewed also on fixed or mobile screens, the best and most immersive way of viewing 3D augmentations is by using AR glasses. Correspondingly, accurate tracking is needed both for positioning augmentations (note that in telepresence this needs to happen over network/distance), and seeing them from a correct viewpoint in the target space. Supporting the same for multiple remote sites and participants causes further complexity, especially if the goal is to support a shared understanding of participant positions (cf. face-to-face meetings).

Table 2 summarizes our exemplary focus on videoconferencing type of photorealistic telepresence approaches (cf. quadrant real human - real space in **Table 1**), with hybrid enhancements based on 3D streaming and XR visualization.

Selected focus: Photorealistic solution based on screen displays		
Challenge	**Approach**	**Solution**
• Shared geometry and depth for natural perception • Participant/user mobility • Avoiding use of HMD/glasses	• Use of (3D) screen displays • Improved support for 3D geometry and depth • Improved support for user mobility • Improved (3D) interaction by (remote) XR	• Support for motion parallax • Support for remote augmentations (remote XR) • Support for viewing XR objects in participant spaces (using mobile displays and/or AR glasses) • Future option for accommodation support

Table 2.
Defining the focus to screen based 3D telepresence solutions.

A simplified hypothesis for our study is that much of the complexity of 3D telepresence solutions can be avoided by aiming at a screen-based solution without a (fully) realistic meeting geometry. An important cue is motion parallax, supported by tracking small user motions and serving with new viewpoints accordingly. Because of this choice, tracking and exchanging user positions for maintaining a unified meeting geometry is omitted, simplifying the solution considerably. Correspondingly, although beneficial in some geometry supporting solutions, the earlier described video-on-demand approach is not needed either.

Despite the demarcation to screen displays, the solution can be enhanced with remote XR functionalities, i.e. by bringing benefits of hybrid approaches to a photorealistic screen based solution. With screen displays, it is also easier to support natural occlusions when compiling remote views (e.g. no need to use XR approaches for displaying remote views around a local participant). Using external (flat) screens is naturally also a solution to avoid (the need of) covering faces by glasses display, i.e. better supporting photorealistic capture of participants.

Correspondingly, the next chapter focuses on describing the above photorealistic approach for 3D telepresence, giving more details on its main challenges and the status of related technical enablers. Most important of those enablers is support for coding and streaming RGB-D data, for which an exemplary implementation is described with some numerical results.

3. 3D telepresence solution using screen displays and supporting XR

3.1 Introduction

In the following chapters, main choices, enablers and components are described for a photorealistic telepresence with screen displays. Features from hybrid approaches are included, e.g. possibility to replace visual captures of a remote participant by an animated avatar. Further, in addition to screen based communication, XR interactions can be supported separately by streaming 3D scanning results between meeting sites, and viewing either locally or remotely produced augmentations e.g. by AR glasses.

3.2 Serving viewpoints by screen displays

Generally, serving moving participants requires views from arbitrary viewpoints. This in turn requires tracking of participant positions and virtual meeting geometry

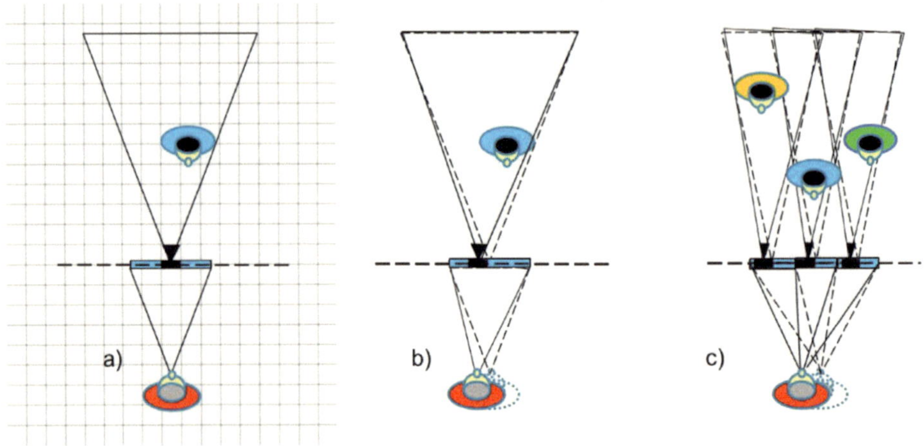

Figure 6.
Supporting motion parallax to a mosaic of 2D or 3D renderings on screen.

in real time. Further, although it may be enough to model a meeting environment in advance, photorealistic 3D capture of participants needs to be made in real time. This in turn requires a setup of multiple 3D sensors and an efficient reconstruction algorithm.

It is however possible to simplify the implementation considerably by relieving from natural geometry requirement. In the minimum, small motion parallax and even natural focus (e.g. using MFPs [16, 17, 25]) can namely be supported without forming and maintaining a virtual meeting geometry between participants. Although with more limitations than with NEDs, also flat screens can support user mobility and consistency of meeting geometries.

By relieving geometry constraints, more freedom for display arrangement and mobility can be achieved. For example, motion parallax can be supported also when compiling remote participants into a video mosaic on a display (a typical situation during video conferencing, as such) and thus to better support 3D cues (**Figure 6**). Note that it may not be that harmful even if all remote participants have their faces oriented towards a local viewer (cf. a "positive Mona Lisa effect", i.e. getting an eye contact even when not being looked at).

In this simplified approach, accurate tracking and delivery of user positions is not needed, and neither is the definition for a unified meeting geometry. Instead, the tracking reduces to a local and rather approximate process of detecting the direction of participant motion, indicating more a viewer's qualitative desire to perceive motion parallax. Further, by satisfying to frontal 3D captures, only one capture sensor is required.

As a result, 3D cues supported by the suggested system are limited to synthesized motion parallax, true eye contact (cf. avoiding the effect of a camera-display offset), and perception of depth. All these are important improvements over existing videoconferencing solutions. As described in [16, 17, 25], supporting natural focus/accommodation is also possible, provided that practical solutions for MFP displays or alike come to market.

3.3 Simplified user tracking and geometry formation

Generally, user tracking and positioning is an important functionality of 3D telepresence solutions. User positioning is required for 1) forming a consistent virtual geometry between participants, and 2) serving a participant with viewpoints complying his/her movements in the defined meeting geometry. A tracking device can be carried by a participant or can measure the person from outside. Visual tracking is commonly assisted by other electronic sensors (IMUs or a like) and by fusing the results for better accuracy.

For a telepresence session, most favorably, a common server makes the formation of a virtual meeting geometry. For this purpose, the server needs participant positions from each telepresence terminal (cf. varying participant positions in **Figure 3c**). Bitrates for delivering 3D positions may be reduced by a suitable coding method, e.g. differential, run-length (RL), variable length coding (VLC), or their combination.

In general, user tracking, from either outside or by wearable sensors, has evolved considerably in recent years. A good solution to provide 6DoF head motion tracking is visual-inertial odometry (VIO), which estimates the relative position and orientation of a moving device in an unknown environment using a camera and motion sensors (https://en.wikipedia.org/wiki/Visual_odometry). A big advantage of VIO is that it can be processed on glasses or HMD without external setups, i.e. sensors, markers, cameras, or lasers set-up throughout the room. A comparison of several VIO approaches is presented in [26].

Note that in our simplified telepresence approach using screen displays, the perception of a consistent geometry between participants is relieved to ease up the implementation. For the screen-based communication, there is no need to derive user positions accurately, nor to deliver them to remote sites. Correspondingly, there is no need to track a camera or cameras for 3D reconstruction either. For supporting motion parallax, rather qualitative detection of user motions is enough, i.e. to detect simply, whether a viewer is moving slightly (e.g. leaning left or right) to perceive a slightly altered view. These small viewpoint changes can be supported locally, e.g. by synthesizing the viewpoints, so that there is no need to deliver captured motions to other participants.

In case the solution is enhanced by the support for seeing augmented objects in a participant's space, the tracking needs to be more wide base and accurate. However, if the support is only for seeing XR objects locally, there is no need to deliver viewer motions to other sites.

3.4 Coding and streaming 3D data

Efficient 3D capture, coding and streaming are important for future 3D telepresence solutions [27–29]. As we introduced in Chapter 2.4, coding and delivery of 3D volume data is not reasonable nor necessary for supporting spatial faithfulness, as a viewer is able to see a 3D environment or content only from one (binocular) viewpoint at a time. This suggests a solution using viewpoint-on-demand approach, which, instead of delivering complete 3D views, serves remote viewers with video-plus-depth (V + D) perspectives from desired viewpoints. A prerequisite of this approach is that user positions are tracked and set into a unified geometry defining (virtual) lines-of-sight between participants.

Luckily, V + D format suggested above serves also well in enhancing screen based telepresence solutions, both with additional 3D cues (motion parallax, and depth, both for stereoscopy or supporting natural focus/accommodation) as well as with XR visualizations and functionalities. Although communication is based on viewing remote participants on screens, a system can also support producing and delivering XR objects, viewed by a local participant's with glasses or by looking through a mobile device.

Using video-plus-depth captures simplifies and eases-up the implementation and reduces bitrates and complexity in data coding and streaming. We applied existing video coding methods supported by FFMPEG for encoding of RGB-D data (e.g. HEVC/X265). A basic challenge is that Kinect type of sensor produces 16 bit/ sample depth values, which are not supported by video coding methods. For that reason, we rounded 16 bit depth values to closest 12 bit integers before coding.

Figure 7 illustrates the pipeline in our experiments. The quality of our video-plus-depth coding and streaming was experimented by comparing direct reconstruction result of a moving RGB-D sensor to the reconstruction made after coding and streaming the data by a HEVC/X265 (FFMPEG) codec. The reconstruction algorithm was the one provided by Open3D. The test sequence was sequence 016 (here denoted as 'Bedroom') from the SceneNN dataset at http://www.scenenn.net, obtained by using Asus Xtion PRO, a Kinect 1 type of depth sensor. The sequence consists of 1364 color and depth frames (captured in about 45 seconds), both in PNG format with 480x640pel per frame.

In **Figure 7**, video and depth sequences were transferred into two video type sequences using the sc. depth blending, modulating the original input video by a linearly weighted depth map and its inverse [25]. This results with two video-like sequences with the partition-of-unity property, meaning that the output sequence is obtained by summing up the modulated (and coded and streamed) video components in the receiver. The coded depth map sequence is obtained from the ratio of luminance(s) for the corresponding pixels. Note that the same approach is typical when forming MFPs for accommodation supportive displays. Here, we omit further details of the coding process and suggest an interested reader to study e.g. the above references.

In our experiment with the above Bedroom sequence, the average bitrate for the original video-plus-depth data from the sensor was 103Mbit/s (RGBD frames in png format, 30fps), and the average bitrate for the coded and streamed data was 567kbit/s, corresponding to about 180:1 compression ratio. Standard (RMS) deviation of the output voxels was 4.2 mm compared to the input ('original') surface, as derived from the reconstructions by the CloudCompare SW (see https:// en.wikipedia.org/wiki/CloudCompare). PSNR was calculated from the differences between corresponding YCbCr pixels of the input and output sequences. Average PSNR was 50.3 dB for the luminance (Y), and 48.6 dB and 55.2 dB for Cb and Cr components. YCbCr format was chosen for being traditionally used in compression research and for better specifying obtained PSNR values. Calculations were made using Matlab (r2018b) functions for format conversions and PSNR. These above numerical results are very good, and when viewing by eye, both the video and the reconstructions appear identical.

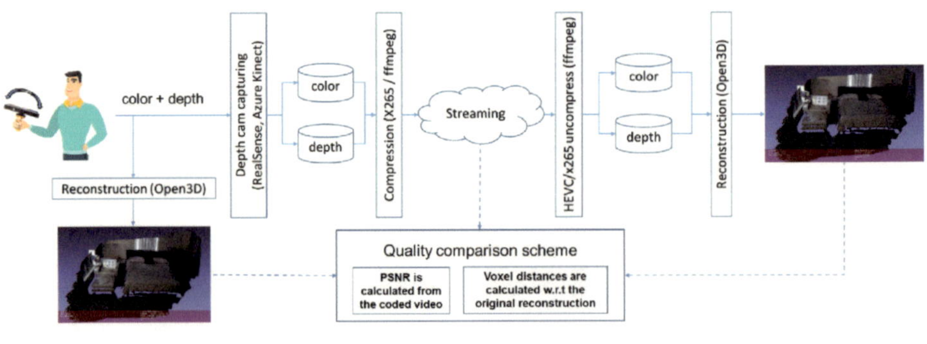

Figure 7.
3D pipeline showing data (color plus depth data from a moving RGB-D) being captured, coded, streamed, and reconstructed in our experiment. Quality comparison (PSNR) is made between original and coded videos, and reconstruction accuracy (RMS distance to the nearest voxel) is measured from 3D reconstructions using original and coded depth images.

Note that the above pipeline is still an offline implementation, using stored files for both input and output data. Correspondingly, there were no real-time limitations in the above simulations. The writers expect to complete also a real-time version of the pipeline by the autumn 2021.

3D streaming solution described above can obviously support also higher resolutions. E.g. with a fourfold resolution to the experiments, the bitrate would remain in the order of 2Mbit/s. As long as there are no better coding methods available, it will be more challenging to support higher pixel dynamics, i.e. more bits/pel (e.g. by depth blending as in our solution).

Generally, the bitrate for video-plus-depth is much less than when streaming multiple-view or volume videos. As a comparison, the approaches used in [7] resulted with the average of 1–2 Gbps transfer rate for a 30 fps stream. According to Qualcomm[1], 6DoF video demands bit-rates in the range of 200 Mbps to 1000 Mbps depending on the end-to-end latency. These figures are just indicative, as the bitrates depend heavily on the used coding scheme and many factors affecting quality (notably resolution used in 3D data capture). Interested readers may find more information from the references given in the beginning of this chapter.

Our simulations on 3D streaming indicate that reconstructing 3D models from coded and streamed video-plus-depth data succeeds with an adequate quality for 3D viewing and reconstruction. Note that the same simple data pipeline enables also various remote support functionalities, including remote 3D analysis based on coded information.

The principle of using compressed data for visualization and analysis is denoted as compress-then-analyze (CTA) approach [30, 31]. According to [31], the opposite analyze-then-compress (ATC) approach may outperform at low bitrates. However, ATC limits a system flexibility, as for example normal viewing of the stream is not possible using only received visual features. Further, ATC fixes the feature selection method at a captured space, limiting applicable approaches for remote analysis. In fact, CTA provides superior flexibility in multipoint settings by enabling for example any analysis approach by multiple remote receivers. According to [30], CTA may also outperform ATC at high bitrates. It is worth noticing, that the referred studies for CTA used jpeg compression for the visual features, which wastes bitrate and lowers quality compared to our efficient spatiotemporal CTA approach.

3.4.1 Reducing the need for streaming by synthesizing viewpoints

Video-plus-depth data format supports synthesizing viewpoints without ordering and streaming new data. This in known from the sc. depth image based rendering (DIBR) approaches for stereoscopic (S3D) TV [15]. Using DIBR, stereoscopic image pairs can be formed in any desired baseline orientation. Viewpoints can also be synthesized to support 3D motion parallax, i.e. any small viewpoint changes around nominal viewpoints from which video and depth images are captured. In a telepresence solution, synthesizing viewpoints can thus be used for both reducing bitrates and avoiding possible latencies of a viewpoint-on-demand approach.

Applicable methods for synthesizing new viewpoints are virtual viewpoint generation in 3D (3D geometry calculations), which are also well supported by graphics processors for speeding up computations. Another way is used in [16, 17], where new viewpoints are formed by simple shifts of MFPs, generated also using video-plus-depth data. The latter approach is good at least if a graphics processor is not available, and if natural accommodation is supported by an MFP approach. Note

[1] https://www.qualcomm.com/media/documents/files/augmented-and-virtual-reality-the-first-wave-of-5g-killer-apps.pdf

that MFPs can also be used for virtual viewpoint generation for normal stereoscopic pairs without the aim for supporting accommodation [16, 17].

4. Enhancements by XR and naturalness

Traditionally, the gold standard for telepresence solutions has been a face-to-face meeting. While mimicking physical encounters over network is technically very challenging, the goal for telepresence solutions has even been raised to exceed the possibilities of physical meetings, referred also to as "beyond being there" [32] and "beyond being aware" [33]. Bill Buxton et al. referred to additional functionalities enhancing face-to-face collaboration as 'groupware' [4]. Correspondingly, also our 3D telepresence solution can and needs to be enhanced with additional functionalities. In our case, many of them are based on XR functionalities, which are discussed in the following.

4.1 Hybrid functionalities for human collaboration

Hybrid functionalities (cf. **Table 1**) combine real and virtual components when rendering and displaying telepresence views. There are two main options, which are described shortly in the following, namely:

1. Replacing a camera captured (and animated) participant view with a virtual avatar (augmented reality option), and

2. Replacing a camera captured participant space with a virtual space (augmented virtuality option).

There are multiple implementations and services using the first approach, e.g. https://remoteface.ai (+ YouTube https://www.youtube.com/watch?v=prpPqwV5Weo). This approach requires either capturing a participant's facial features in order to animate the avatar, or in the minimum, capturing a participant's speech to estimate underlying facial muscle movements and corresponding animation parameters.

The second approach was already illustrated in Chapter 2.5, where a real-time captured human is rendered into a typically in-advance modeled virtual space (cf. e.g. **Figure 4**). Note that a virtual space may even enable a remote participant to make virtual visits to that space (i.e. seeing to it from widely varying viewpoints) – in particular, if the viewing is supported by glasses display.

Note that using glasses for viewing makes the interaction easily nonsymmetrical or even one-way, as the glasses prevent either capturing facial movements of their wearer, or seeing his/her face and eyes. Somewhat working solutions to avoid this have however been described in literature, based e.g. on real-time manipulation of facial areas [7].

4.2 Remote XR support functionalities

When developing 3D telepresence solutions, we are particularly interested in supporting remote XR functionalities. There are two main approaches for doing it. The first approach requires only delivering of images or video to the remote site(s), and coding and streaming is supported straightforwardly by existing video coding methods. However, better support for remote interaction is provided by coding data from a depth sensor, and after streaming the data, making 3D reconstruction at a remote site. Algorithms used for local reconstructions are applicable also for remote reconstructions.

Thus, in addition to better 3D perception, video-plus-depth data supports also forming (or copying) 3D reconstructions at remote sites. These reconstructions, which we have denoted as Visual Twins, can support various 3D remote support functionalities, e.g. 3D monitoring, control, and analysis, as well as remote augmentation with visualizations and instructions. As described in Chapter 3.4, this is feasible by applying existing coding methods.

We have tested both video-based (sc. Ad-hoc AR) and video-plus-depth based (sc. Visual Twin) approaches, and they are described in more detail in the following:

1. Local 3D reconstruction, pointed remotely for positioning augmentations (Ad-hoc AR)

In this option (**Figure 8**), a 3D reconstruction is made in a local space using e.g. an RGB-D sensor carried by a moving person or a robot. The orientation of each RGB-image is derived in a normal way in the reconstruction process (e.g. using SLAM [34] and TSDF [35]), and stored locally with the image ID (e.g. a simple timestamp). The images are coded and streamed separately (e.g. following a manual selection) or as a sequence to a remote space. In the remote space, a person selects a point (pixel) in an image to show an augmentation, and messages back the image ID, target pixel coordinates, and data (or ID, if stored on a common server) of the AR object.

At the local site, the image's orientation w.r.t. 3D reconstruction is fetched from the local memory. The point to show the augmentation is obtained by ray-tracing through the defined pixel to the known orientation. Ray-tracing defines a 3D surface point on the 3D reconstruction (and the space), and enables local participant(s) to see the augmentation from various directions. **Figure 8** illustrates the process.

2. Remote 3D reconstruction using streamed depth sensor data (Visual Twin)

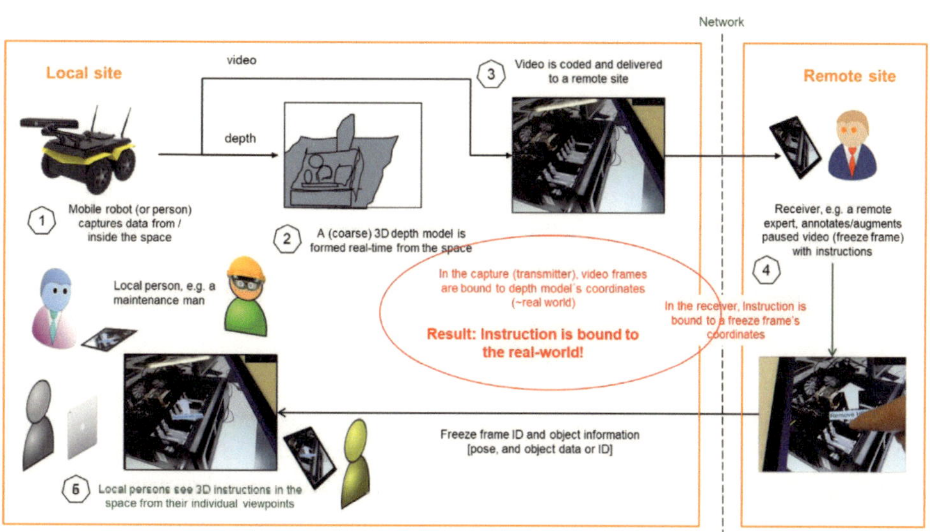

Figure 8.
Ad-hoc AR, enabling remote augmentation of a locally reconstructed space.

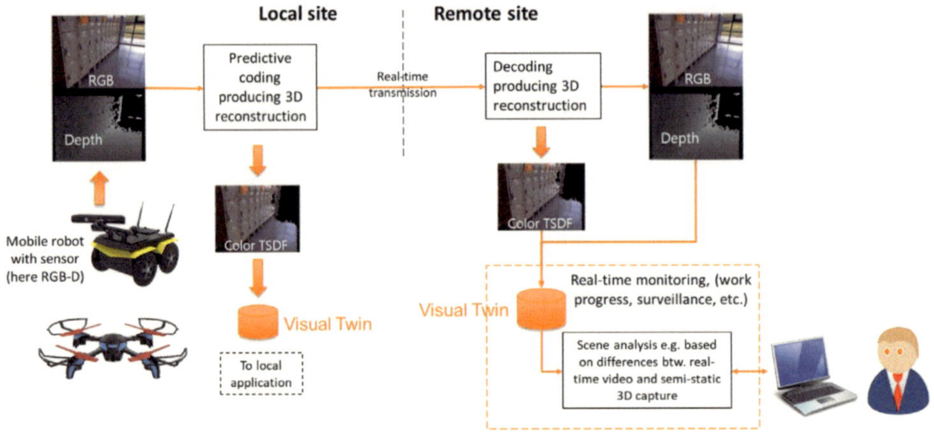

Figure 9.
Principle of visual twin, here used for remote monitoring & control.

In this approach (**Figure 9**), data for 3D reconstruction (e.g. RGB-D data) is coded and streamed over a network, and the reconstruction is made using the data decoded at the remote site. The solution described in Chapter 3.4 for the coding and streaming of video-plus-depth data supports directly this approach when made in real time (implementation for this is soon completed by VTT).

Both of the approaches have been implemented and demonstrated by VTT, and are suitable for enhancing our 3D telepresence solution with XR functionalities. The first ray-tracing based option is simpler, but being based on images/video only, does not allow viewing the data in 3D at the remote site. Note that receiving video-plus-depth data enables also the ray-tracing approach, meaning that a combination of the approaches is also possible.

4.3 Supporting lines-of-sight

A straightforward way to enhance the described solution is to support virtual lines-of-sight between participants in the same way as in the describe Hydra approach, i.e. by using an own screen based terminal for each of the (maximum number of) remote participants (cf. Chapter 2.1). This means full mesh connections between peers, but the total bitrate may remain low due to each of the streams being a low bitrate video-plus-depth stream. This enhancement would also enable also earlier referred options for increasing user (plus meeting site) mobility e.g. by grid based geometries and interactive landscapes (cf. Chapter 2.2). The downside of this approach is the increase in the complexity for a meeting setup, and support for only a limited number of participants due to the lack of space around a participant. This version may however be justified in special cases where better spatial and gaze awareness between interactive participants is particularly important.

4.4 Natural eye-focus/accommodation

Stereoscopic rendering is a further means to improve 3D perception. However, stereoscopic viewing detaches natural eye focus (accommodation) and convergence distances. The resulting vergence-accommodation conflict (VAC) causes discomfort and nausea, and restricts a person's willingness to view stereoscopic content [6, 25, 36]. A related conflict in monocular viewing is the sc. focal rivalry (FR) [37]. In addition to VAC and FR, there are various other error types caused by the wrong blur of rendered scene components, e.g. all-in-focus virtual objects, or real objects blurred by camera optics. These types of distortions appear especially when combining content for augmented scenes. Without careful considerations, these distortions continue hindering the quality and acceptance of XR functionalities.

As described in Chapter 2.6, unnatural occlusions are another common type of distortion, which occurs especially in XR, when combining optical views (cf. OST glasses), photorealistic captures, and 3D modeled objects and spaces. In addition to occlusions, supporting natural focus is a big challenge in display design and manufacture, and has not yet been properly solved for either external screens or near-eye displays. Natural focus requires that a content is rendered into a 3D volume, instead of one or few display surfaces. There are volume displays aiming to this, but generally they suffer e.g. from the lack of occlusions, colors and brightness.

There are various means to support natural focus, including for example light field rendering [38]. Supporting occlusions in a light field display has been studied e.g. in [39], and occlusions for its simple variant multiple-focal-plane (MFP) display in [25]. MFP-rendering is essentially light field rendering to one viewpoint, and as such fits well to the above introduced viewpoint-on-demand approach. The writers have studied various ways of forming and rendering MFPs, and have even made own designs for MFP glasses [40].

Implementation of MFP or other accommodative glasses is however very challenging, as for example the efforts by Magic Leap, Inc. has us learned. However, as glasses are superior in supporting immersion and mobility, we expect that major technical problems will be solved, and high quality accommodation support will be available within about a decade. Note that their (eventual) emergence will in general mean a big change to the ways visual information is captured, processed and displayed.

5. Conclusions and acknowledgements

For using a photorealistic approach in 3D telepresence, the biggest challenges are the accuracy and cost of acquiring, delivering, and displaying 3D captures. Glasses based approaches are attractive due to their ability to support user mobility, immersion, and in future even natural focus/accommodation in 3D perception. Glasses can provide also sensors for 3D capture and user positioning. However, natural occlusions and accommodation are hard to support, and likely it will still take many years before affordable and good-enough glasses are available.

In this article, in addition to reviewing the general status and approaches for 3D telepresence, we proposed a simplified approach for spatially faithful telepresence, based on 3D screens and low bitrate streaming of video-plus-depth data. Screen displays are cheap, and can support 3D cues and mobility better than in existing teleconferencing solutions. External displays are easy to view and are a good option for improved 3D telepresence solutions. An important enabler for a simplified solution is efficient coding and streaming of RGB-D data, for which an exemplary implementation was presented with some simulation results. Described means for 3D capture and streaming support also XR functionalities, which in addition to viewing XR content on screens, may also be used with glasses or mobile displays. Features of the introduced solution can be summarized as follows:

- Photorealism with improved support for 3D cues

- Screen displays are preferred due to not obstructing faces

- Gracefully compromising mutual (real-world like) geometry for more freedom in display placement
- Supporting low bitrates by video-plus-depth (V + D) streaming
- Delivering enough 3D data for supporting remote XR functionalities
- A participant can see augmentations in his/her environment by using a mobile device (a "magic lens") or AR glasses
- Hybrid approaches can be supported, with options for:
    1. replacing a participant capture by an avatar, and
    2. replacing a captured meeting environment by a virtual space
- Easy to be modified e.g. for emerging MFP displays supporting natural eye focus

Choosing screen displays implies both limitations and benefits to the system. With one or few screen displays, the perception of a meeting geometry is not same between participants. We decided to rely on the benefits of an exaggerated perception for eye contact by collecting all remote participant renderings into a grid on (nominally) one screen display, and providing viewers with perception for depth and motion parallax.

The key enabler for the improved system is simple: support for real-time capture, coding and streaming of video-plus-depth data from the RGB-D sensor of a telepresence terminal. This choice enables low bitrates, depth perception, and support for small viewpoint changes by user motions. Currently, the most feasible option for a display is a stereoscopic (S3D) display, but it can be replaced by an accommodation supportive display as soon they come available.

Our basic assumption is that each participant has his/her own telepresence terminal in the same way as PCs and laptops are currently used in videoconferencing. Correspondingly, from appearance, the new solution does not differ too much from current videoconferencing systems. For better supporting spatial faithfulness and gaze awareness, a more complicated system with several cameras and 3D screen displays may be used (i.e. applying Hydra and Viewport type of approaches introduced in Chapter 2.1). In parallel to viewing remote participants on screens, AR glasses can be used for viewing augmented objects inside a local space.

Although we have tested only some important prerequisites of the suggested solution, we have now a good knowledge and plan for its full implementation. We hope that this article raises its readers a general interest to the status, challenges, and possibilities of 3D telepresence, as well as a specific interest to develop the described approaches and ideas even further.

We want to express our gratitude to VTT Technical Research Centre of Finland for giving us the opportunity to work on 3D telepresence and related topics. Thanks also for InterDigital Inc. for challenging us to make new inventions in this area.

DOI: http://dx.doi.org/10.5772/intechopen.99271

Author details

Seppo Valli*, Mika Hakkarainen and Pekka Siltanen
VTT Technical Research Centre of Finland Ltd., Espoo, Finland

*Address all correspondence to: seppo.valli@vtt.fi

References

[1] Nguyen D, Canny J. MultiView: Spatially faithful group video conferencing. CHI 2005 Technol Safety, Community Conf Proc - Conf Hum Factors Comput Syst. 2005;799-808.

[2] Monk AF, Gale C. A Look Is Worth a Thousand Words: Full Gaze Awareness in Video-Mediated Conversation. Discourse Process [Internet]. 2002 May 1;33(3):257-78. Available from: https://doi.org/10.1207/S15326950DP3303_4

[3] Siltanen P, Valli S. Gaze-aware video conferencing application for multiparty collaboration. In: 2013 International Conference on Engineering, Technology and Innovation, ICE 2013 and IEEE International Technology Management Conference, ITMC 2013. 2015.

[4] Buxton W, Sellen A, Sheasby M. Interfaces for multiparty videoconferencing. In K. Finn, A. Sellen & S. Wilber (Eds.). Video Mediated Communication. Hillsdale, N.J.: Erlbaum. 1997;385-400.

[5] Zhang C, Cai Q, Chou PA, Zhang Z, Martin-Brualla R. Viewport: A distributed, immersive teleconferencing system with infrared dot pattern. IEEE Multimed. 2013;20(1):17-27.

[6] Kramida G. Resolving the Vergence-Accommodation Conflict in Head-Mounted Displays. IEEE Trans Vis Comput Graph. 2016;22(7):1912-31.

[7] Orts-Escolano S, Kim D, Cai Q, Rhemann C, Davidson P, Chou P, et al. Holoportation: Virtual 3D teleportation in real-time. UIST 2016 - Proc 29th Annu Symp User Interface Softw Technol. 2016;741-54.

[8] Valli S, Siltanen P, inventors. Spatially faithful telepresence supporting varying geometries and moving users. WO18226508 A1. PCMS Holdings, Inc.;2018 Dec 13.

[9] Valli S, inventor. System and method for augmented reality multi-view telepresence. WO17030985 A1. PCMS Holdings, Inc.;2017 Feb 23.

[10] Siltanen P, Valli S, inventors. System and method for spatial interaction using automatically positioned cameras. WO18005235 A1. PCMS Holdings, Inc.;2018 Jan 4.

[11] Smolic A, Mueller K, Merkle P, Kauff P, Wiegand T. An Overview of Available and Emerging 3D Video Formats and Depth Enhanced Stereo as Efficient Generic Solution. In: Proceedings of the 27th Conference on Picture Coding Symposium. IEEE Press; 2009. p. 389-392. (PCS'09).

[12] Chen Y, Vetro A. Next-Generation 3D Formats with Depth Map Support. IEEE Multimed. 2014;21(2):90-4.

[13] Valli S, Siltanen P, inventors. Apparatus and method for supporting interactive augmented reality functionalities. WO2017172528A1. PCMS Holdings, Inc.;2017 May 10.

[14] Valli S, Siltanen P, inventors. System and method for supporting synchronous and asynchronous augmented reality functionalities. WO17177019A1. PCMS Holdings, Inc.;2019 Sep 26.

[15] Debono, C. J., S. Faria, Luis F. R. Lucas and Nuno M. M. Rodrigues. Depth Map Coding for 3DTV Applications; 2017.

[16] Valli S, Siltanen P, inventors. Method and system for forming extended focal planes for large viewpoint changes. WO20009922A1. PCMS Holdings, Inc.;2020 January 9.

[17] Valli S, Siltanen P, inventors. Multifocal plane based method to produce stereoscopic viewpoints in a DIBR system (MFP-DIBR).

WO19183211A1. PCMS Holdings, Inc.;2019 Sep 26.

[18] Hornung A, Wurm KM, Bennewitz M, Stachniss C, Burgard W. OctoMap: an efficient probabilistic 3D mapping framework based on octrees. Auton Robots [Internet]. 2013;34(3):189-206. Available from: https://doi.org/10.1007/s10514-012-9321-0

[19] Frueh C, Sud A, Kwatra V. Headset Removal for Virtual and Mixed Reality. In: ACM SIGGRAPH 2017 Talks [Internet]. New York, NY, USA: Association for Computing Machinery; 2017. (SIGGRAPH '17). Available from: https://doi.org/10.1145/3084363.3085083

[20] Milgram P, Kishino F. A Taxonomy of Mixed Reality Visual Displays. IEICE Trans Inf Syst. 1994;77:1321-9.

[21] Hu X, Hua H. Design and Assessment of a Depth-Fused Multi-Focal-PlaneDisplay Prototype. J Disp Technol [Internet]. 2014;10(4):308-16. Available from: http://jdt.osa.org/abstract.cfm?URI=jdt-10-4-308

[22] Koulieris GA, Akşit K, Stengel M, Mantiuk RK, Mania K, Richardt C. Near-Eye Display and Tracking Technologies for Virtual and Augmented Reality. Comput Graph Forum [Internet]. 2019 May 1;38(2):493-519. Available from: https://doi.org/10.1111/cgf.13654

[23] Kiyokawa K, Billinghurst M, Campbell B, Woods E. An occlusion capable optical see-through head mount display for supporting co-located collaboration. In: The Second IEEE and ACM International Symposium on Mixed and Augmented Reality, 2003 Proceedings. 2003. p. 133-41.

[24] Kantonen T, Woodward C, Katz N. Mixed reality in virtual world teleconferencing. In: 2010 IEEE Virtual Reality Conference (VR). 2010. p. 179-82.

[25] Akeley K, Watt SJ, Girshick AR, Banks MS. A Stereo Display Prototype with Multiple Focal Distances. ACM Trans Graph [Internet]. 2004;23(3):804-813. Available from: https://doi.org/10.1145/1015706.1015804

[26] Pfrommer B, Sanket N, Daniilidis K, Cleveland J. PennCOSYVIO: A challenging Visual Inertial Odometry benchmark. In: 2017 IEEE International Conference on Robotics and Automation (ICRA). 2017. p. 3847-54.

[27] Bannò F, Gasparello PS, Tecchia F, Bergamasco M. Real-Time Compression of Depth Streams through Meshification and Valence-Based Encoding. In: Proceedings of the 11th ACM SIGGRAPH International Conference on Virtual-Reality Continuum and Its Applications in Industry [Internet]. New York, NY, USA: Association for Computing Machinery; 2012. p. 263-270. (VRCAI '12). Available from: https://doi.org/10.1145/2407516.2407579

[28] Fu J, Miao D, Yu W, Wang S, Lu Y, Li S. Kinect-Like Depth Data Compression. IEEE Trans Multimed. 2013;15(6):1340-52.

[29] Merkle P, Müller K, Wiegand T. 3D video: acquisition, coding, and display. IEEE Trans Consum Electron. 2010;56(2):946-50.

[30] Baroffio L, Cesana M, Redondi A, Tagliasacchi M, Tubaro S. Coding visual features extracted from video sequences. IEEE Trans Image Process. 2014;23(5):2262-76.

[31] Redondi A, Baroffio L, Bianchi L, Cesana M, Tagliasacchi M. Compress-then-analyze versus analyze-then-compress: What is best in visual sensor networks? IEEE Trans Mob Comput. 2016;15(12):3000-13.

[32] Hollan J, Stornetta S. Beyond being there. Conf Hum Factors Comput Syst - Proc. 1992;119-25.

[33] Carroll JM, Rosson MB, Farooq U, Xiao L. Beyond being aware. Inf Organ [Internet]. 2009;19(3):162-85. Available from: http://dx.doi.org/10.1016/j.infoandorg.2009.04.004

[34] Izadi S, Kim D, Hilliges O, Molyneaux D, Newcombe R, Kohli P, et al. KinectFusion: Real-time 3D reconstruction and interaction using a moving depth camera. UIST'11 - Proc 24th Annu ACM Symp User Interface Softw Technol. 2011;559-68.

[35] Saputra MRU, Markham A, Trigoni N. Visual SLAM and Structure from Motion in Dynamic Environments: A Survey. ACM Comput Surv [Internet]. 2018;51(2). Available from: https://doi.org/10.1145/3177853

[36] Hoffman DM, Girshick AR, Akeley K, Banks MS. Vergence-accommodation conflicts hinder visual performance and cause visual fatigue. J Vis. 2008;8(3):1-30.

[37] Oshima K, Moser KR, Rompapas DC, Swan JE, Ikeda S, Yamamoto G, et al. SharpView: Improved clarity of defocussed content on optical see-through head-mounted displays. In: 2016 IEEE Virtual Reality (VR). 2016. p. 253-4.

[38] Iwane T. Light field display and 3D image reconstruction. In: ProcSPIE [Internet]. 2016. Available from: https://doi.org/10.1117/12.2227081

[39] Wang T-C, Efros AA, Ramamoorthi R. Occlusion-Aware Depth Estimation Using Light-Field Cameras. In: 2015 IEEE International Conference on Computer Vision (ICCV). 2015. p. 3487-95.

[40] Valli S, Siltanen P, inventors. Multi-focal planes with varying positions. WO19143688A1. PCMS Holdings, Inc.;2019 Jul 25.

Chapter 8

Using Augmented Reality in Different BIM Workflows

Angela Cristina De Hugo Silva, Metod Gaber and Matevž Dolenc

Abstract

Building Information Modeling is an increasingly common process for managing the entire lifecycle of a building - from design and planning, through the construction phase, to operation and maintenance. The result of this process is a building information model with all the generated data and information about the construction process that can be used in a variety of different end-user scenarios. One such use of the model is in a number of different augmented reality applications. Augmented reality technology is being used to bridge the gap between the digital and real worlds and is rapidly becoming an essential part of modern building data modeling design workflows. The chapter provides an overview of building data modeling and the current state of the art in the use of augmented reality in various user scenarios of building data modeling and explores various challenges that need to be addressed for the adoption of augmented reality technology in architecture, engineering, and construction in general.

Keywords: augmented reality, computer integrated engineering, civil engineering, project documentation, construction, building information modeling, BIM

1. Introduction

According to [1], Building Information Modeling (BIM) [2] is a set of technologies, processes and policies enabling multiple stakeholders to collaboratively design, construct and operate a Facility in virtual space. The result of a BIM process is a building information model that a shared digital representation of a built asset to facilitate design, construction and operation processes to form a reliable basis for decisions. The term BIM continues to evolve over the years and is thus best understood as an 'expression of digital innovation' across the construction industry and the overall built environment and is an increasingly common process for managing the entire lifecycle of a building - from design and planning, through the construction phase, to operation and maintenance. As noted by D. Richard and C. Harty [3] highly structured and semantically rich 3D geometry information is in practice used on physical 2D medium, e.g. paper or digital displays. So at the end, basic concepts remain the same even though we now have access to much richer information with BIM models.

As not much has changed on the conceptual model of translating virtual 3D plans into real world structures, engineers still have to rely on their knowledge,

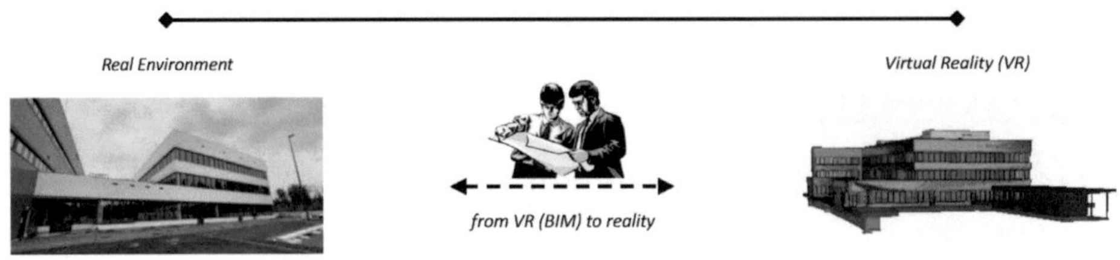

Figure 1.
From virtual reality, e.g. BIM model, to real environment.

experience and spatial awareness to map these virtual plans, that may include 2D drawings (plans, elevations and sections) or 3D models (viewed on a 2D display medium) into real environment (see also **Figure 1**). This challenge has been recognized by different researchers, including P. S. Duston and H. Shin [4] that proposed this mapping process be facilitated by mixing virtual information with the real environment. A technology that enables mixing digital information with the real world environment objects and spaces is called augmented reality (AR) [5].

AR Systems generally consist of three main phases: Data Phase, Computation, and Presentation. The data phase is primarily concerned with the creation, curation, and formatting of data. In the context of BIM related workflows, this means designing a BIM model so that it can be used as a source for augmenting reality. Merging virtual data (3D geometry, specific element information, etc.) and real environments is a computationally intensive phase and can take place on the mobile device or on a remote server. Finally, the display of the mixed visualizations can be done on handheld mobile devices such as smartphones and tablet computers and (2) head mounted devices. Several AR systems already exist; however, they are mainly used to display proprietary models prepared using specialized custom software, including the software that was used for this research [6].

The chapter provides an overview of building data modeling and the current state of the art in the use of augmented reality in various user scenarios of building data modeling. Specifically, it describes various challenges that need to be addressed as well as a spectrum of different end-user scenarios and use-cases for use of AR technology in architecture, engineering, and construction (AEC). As part of the conclusions, the SWOT analysis of using augmented reality system in to context of BIM is also provided.

2. Building information modeling

The roots of Building Information Modeling (BIM) can be traced back to the first ideas on how to use the concept of product models using various media in architectural designs [7], as reported by Russell and Elger [8] regarding the introduction of BIM into the AEC market. And following this evolution, the beginning of the Industry Alliance for Interoperability (IAI) in 1995, and from experiences established a standard for describing buildings, which would allow the exchange of information about buildings without the loss of their semantic information [8]. This working format is called Industry Foundation Classes (IFC) [9], having been published for the first time in 1997.

Following the definition of BIM by the US National Building Information Modeling Standard (NIBS 2012), BIM digitally represents the physical and functional characteristics of a building, enabling the collaboration of different stakeholders throughout the building's lifecycle to enter, update or modify information in the BIM model. BIM is always evolving, seeking to optimize technology according to the complexity of the processes applied in civil construction and the construction community is looking to innovate with BIM through specific workflow tools such as VR and AR that are being applied directly to solve real-world problems as installation verification and estimating [10].

The essence of building information modeling is the ability to add useful information [10] using BIM models that are more than geometric representations; therefore, they can be viewed in various dimensions, from 3D (design planning), 4D (scheduling), 5D (costing), 6D (life cycle information) and 7D (facility management) [11] (**Figure 2**) and, the data models used differ from each other according to the schema used both to organize the data and the schema language to transport the data.

The IFC data model it is an open and platform-neutral file format, facilitating interoperability in the architecture, engineering and construction industries, being used through specialized BIM programs and therefore with its own platforms such as Revit, ArchiCAD, Navisworks, Bentley, among many others. Since IFC is an open standard it enabled development of different translators from one data format to another, for example from IFC to XML [2].

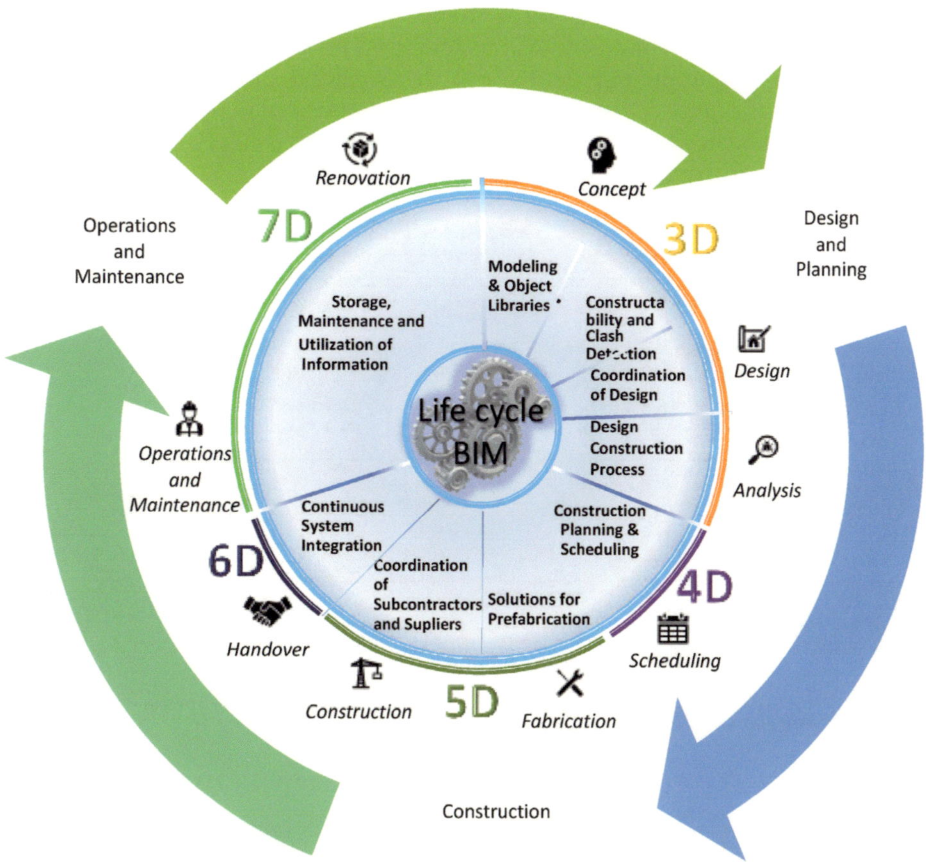

Figure 2.
Building information modeling enables communication, collaboration and visualization of BIM models throughout the building life cycle; from design and planning, through construction phase, and finally operations and maintenance.

3. Augmented reality

Current virtual reality technologies are based on ideas constructed and reported from the 1960s and possibly earlier, such as the iconic Ivan Sutherland who in 1968 created the first head-mounted display that rendered simple wireframe models for pose change of the viewer [12]. It was through the foundations of this invention and with technological innovations and other evolutions that we now call Virtual Reality (VR), Augmented Reality (AR) and Mixed Reality (MR) (**Figure 3**) [13].

With the concept of VR conceived more than fifty years ago, when the first immersive human-computer interaction (HCI) mock-up called "Human-Machine Graphic Communication System" was invented [12] the formal term of VR was placed in 1989 on the RV continuum based on Milgram's taxonomy where VR represents the creation of a virtual reality or. environment in which the user can enter this scene with the feeling of being in the "real" world, with limited level of "realism" such as visual and sound effects [14].

According to this definition, AR is an environment where additional data generated by the computer is fed into the user's view of a real scene [15]. With AR, users can access, visualize and interact with complex information in the context of the real environment, or in other words, computer-generated elements are added to the seen reality. Since the entry requirements for a AR capable device are relatively low, many of today's smart devices (phones, tablets) are suitable for AR use.

With the growth and maturation of technologies in AR, applications end up becoming more viable and popular for both the education, design, manufacturing, construction and entertainment sectors, becoming potential in helping to improve existing technologies and, with that, can promote a better quality of life mainly for people with physical and/or mobility limitations.

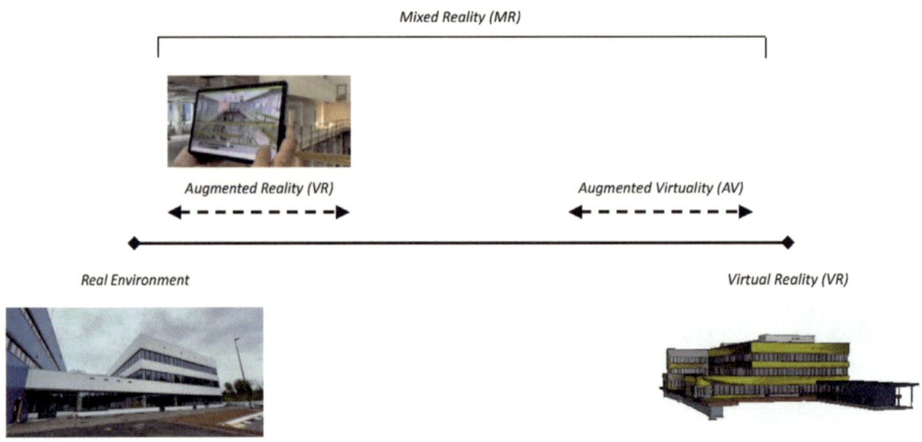

Figure 3.
Reality-Virtuality continuum [13].

4. Augmented reality in BIM

In the context of augmented reality and BIM we can define augmented reality can as system in which BIM model is used to augment real environment. An AR enhanced BIM modeler could also be envisioned where BIM model information is edited in an AR environment. AR is currently being considered by scholars and practitioners in the area of knowledge, as a "New Age" of information through advanced technology due to the positive effects of its technological potential that its correct use offers to the construction industry, especially in the construction phase, providing significantly affects the efficiency of projects, quality, health and safety [16], and consequently the project cost and duration, positively [17].

In Section 2. we already established BIM as a type of a "central hub" to all data and information related to a specific projects – this of course includes 3D models necessary for AR applications as well as non-geometrical information that could also be linked and accessed in AR applications.

But combining real world objects with virtual ones can be challenging. Researchers Bajura M and Neumann U [18] identified four main challenges to be addressed: (1) the origin of the tracking system is not aligned with the world coordinate system; (2) the transformation from source to object is not accurate; (3) the position of the virtual camera is not correct – usually related to inertial and motion-based sensors errors; and (4) virtual camera-to-image mapping does not accurately model the real camera.

To organize the data processing steps for the final BIM visualization in AR, Williams et al. [19] describes a generalized three step workflow (similar architecture is also proposed by Meža et al. [20]): (1) First, it is necessary to generate geospatial properties for each BIM object where mobile AR application uses geolocation to identify the user's position and, consequently, be able to produce information related to the viewed object; (2) Second, from the moment that the information displayed in mobile AR refers to a user's position, several points surveyed within BIM also need to be identified. These points will represent where users can stay in a physical location and perform mobile AR tasks with BIM data sets related to that location; and (3) Third, for BIM to be usable in a general mobile AR environment, the geometry and property data set needs to be separated into two exchange formats.

5. Use-case scenarios

In the context of BIM workflows, the use of augmented reality is usually associated with three basic use-case scenarios [21]: (1) design phase - review of proposed solution, (2) construction phase - monitoring of construction progress [22] and (3) operation phase – building maintenance [4]. But as identified by C. Woodward and M. Hakkarainen [23] many other relevant use-case scenarios should be considered, including layout optimization, excavation, positioning, inspection, coordination, supervision, commenting, etc.

In the following subsections, three different representative use cases are described. Two of them are based on infrastructure projects (railway/tunnel, bridge) and the last one is based on an office building. In all presented use cases, a BIM specific mobile application [6] has been used to access BIM models as well as to use the integrated AR solution. The mobile apps use a BIM shared data environment solution that provides a central hub for all person-to-person communication, data sharing, etc.

5.1 Use-case 1: infrastructure (design and planning phase)

The project used for this use-case is a construction of the second railway line from Divača to Koper that includes construction of more than 17.4 km of access roads with various structures serving as service lines for the construction of tunnels, retaining walls and a bridge approximately 35 m long. It is one of the largest infrastructure project in Slovenia.

The first use-case clearly falls into the design and planning phase of the full life cycle using AR for visualization of the construction site organization as well as visualization of the completed structure in real environment. One of the main goals for the use-case was also to examine how AR technology and its implementation in the selected mobile application can be used for an infrastructure projects that are defined primarily with long distances.

We used a tablet computer with augmented reality application to test the use-case in two different locations that are a few kilometers apart. However, both are georeferenced in a 2D model on the tablet (**Figure 4**) and correctly link the required information in the 3D model.

Its integration through AR technology to visualize interactive 3D models on site [24, 25] (**Figure 5**). As seen in the images bellow, it was possible, for example, to check the tunnel portal and MEP installations, railway track to be build, as well as to get the general scale and complexity of the construction site.

Key takeaways: (1) it is absolutely essential that all BIM models are correctly georeferenced for use by AR system, (2) use of tablets/phones can be challenging in bright environments, (3) difficult to orient yourself in the field and correctly position and scale BIM model, (4) can be used for general understanding of future construction site, and (5) determining geolocation can be challenging and it is depended on geographical characteristics of the construction area.

5.2 Use-case 2: infrastructure (operation and maintenance phase)

The project used for the second use case is an already completed infrastructure project - a new bridge (**Figure 6**) over the Savinja River in Marija Gradec, near Laško. The bridge is 123 m long and up to 12 m high and designed as an anchored containment structure. The bridge construction started in mid-July 2019 and was opened for traffic in February 2021.

The use case clearly addresses the use of AR in the operation and maintenance phase (7D, see also **Figure 2**) of the life cycle [26, 27]. BIM model (3D geometry only) was available and correctly georeferenced via the mobile platform used [6]. This made it possible to obtain real-time architectural visual interaction and communication. Another important point for the choice of this site as a case study is precisely the counterpoint that apart from the 3D model BIM in IFC format as basic information, no other accessible database is available, which imposes major limitations for its use and visualization in augmented reality. Overall, it is possible to check the 3D model BIM and compare the same coordinates on Google Maps as shown in **Figure 7**.

Key takeaways: (1) it is essential for the of AR system, that BIM model includes additional information to basic 3D geometry, (2) positioning/scaling of BIM model in AR system possible due to available reference points, (3) difficult to use tablets/smart phones in bright sunlight.

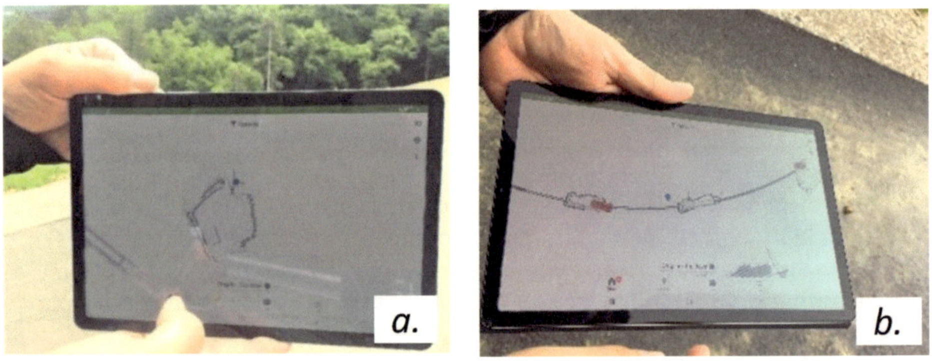

Figure 4.
The stand point 1 (a) and the stand point 2 (b).

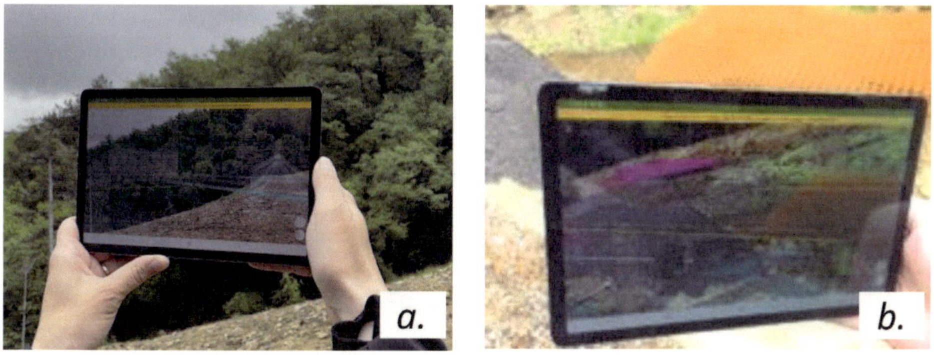

Figure 5.
Visualizing (a) a railway track and (b) a portal to a tunnel.

Figure 6.
The Marija Gradec bridge.

Figure 7.
The geoposition of the bridge in Google maps (left) and BIM model (right).

5.3 Use-case 3: office building

The construction of the industrial complex "Industrial and commercial building Iskra Mehanizmi Brnik" (**Figure 8**) is structurally divided into 3 main blocks: the commercial part, which consists of a monolithic reinforced concrete structure and two production and storage units made of prefabricated reinforced concrete.

The use case focuses on the Iskra Mehanizmi office building (**Figure 9**). It is a complex structure that is in the final stages of construction. The example can be

Figure 8.
Iskra Mehanizmi Brnik industrial complex.

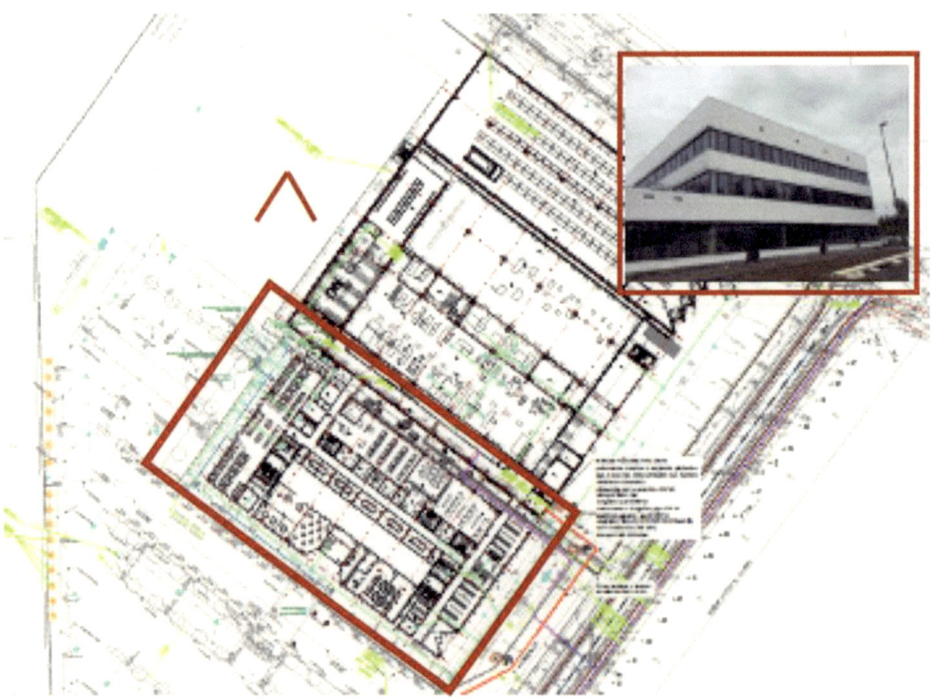

Figure 9.
The site situation/implementation plan. Top right is the newly constructed Iskra Mehanizmi office building.

used to demonstrate the use of AR technology at different lifecycle stages (see also **Figure 2**). A complete common data environment was available for the project, which contained various 3D models BIM with all associated information, allowing the use of AR.

The model BIM supports geometric and non-geometric design information such as location through geo-referencing (3D) and technical specifications (4D) from door and window manufacturers as well as maintenance information (5D) (**Figure 10a**). The convergence between the 3D model BIM and the AR (**Figure 10b**) is practically done through three main components: the 3D model BIM itself, the whole readout and the transformation of the data for interpretation in augmented reality through an appropriate application on a mobile device, in this case a tablet was used. To enable visualization on the AR platform, the 3D model BIM sends and receives information BIM, such as installation adjustment requirements and work plan review, allowing the user to interact with the 3D model BIM and other members of the

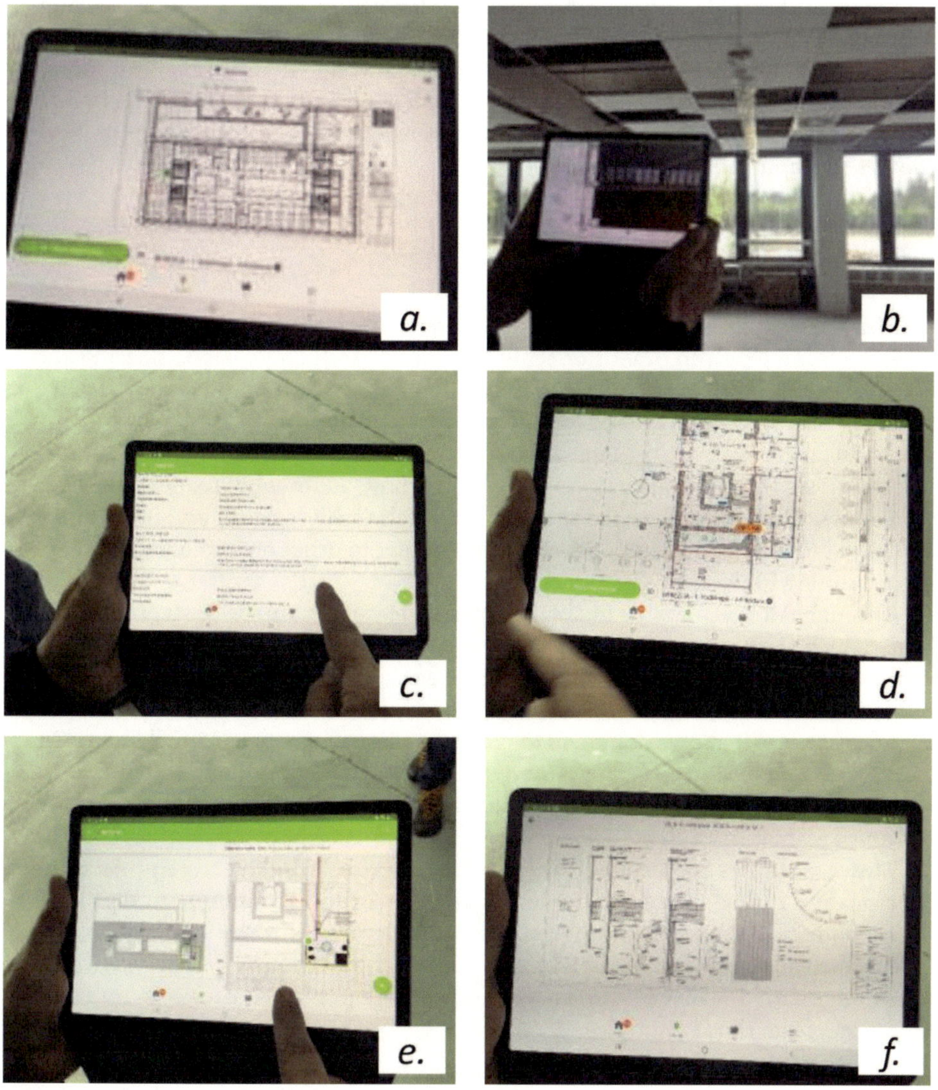

Figure 10.
Using an AR system - (a) location used for geospatial reference, (b) positioning/orienting/scaling the BIM model, (c) accessing associated tasks for current location, (d) information data request, (e) checking specific information of a BIM object, (f) accessing construction details.

project team in real time (**Figure 10c**). BIM models are created based on 2D design drawings and with information included in the design specifications according to the client's requirements. In this case, there is a marker (orange rectangle) that refers to the requested information (**Figure 10d**). After clicking on the orange rectangle, a new window immediately opens with the details of the information in a 2D drawing with comments about the location, the type of installations performed, and what needs to be done to complete the jobs (**Figure 10e**). The levels of construction details included in the same information can be structural, architectural, MEP installations or others required to complete the project (**Figure 10f**).

One of the main objectives for the use case was also to test the use of AR inside a building, where it is usually impossible to collect GPS signals for geolocation of the stand position. The mobile solution used in the example, AR [6], uses a manual geolocation method where the user (1) selects an approximate position in the 2D plane and (2) rotates and scales the 3D model based on reference points (windows, doors, columns, etc.).

Key takeaways: (1) determining geolocation within a building can be challenging and different AR may take different approaches to solve this, (2) positioning and scaling a BIM model in the AR system is possible due to the available reference points, (3) BIM models are usually more complete for buildings compared to infrastructure objects (bridges, tunnels, etc.), therefore the AR system can make better use of the available information.

6. Conclusions

This paper mainly focuses on the major application scenarios mentioned by various researchers and practitioners in the field of architecture, engineering and construction. The advances in the digitization of all processes and workflows in the AEC industry in general and beyond with the wide adoption of BIM as a common methodology for managing large construction projects enables the use of advanced information communication technology (including, augmented reality) in AEC workflows. It is not emphasized that the approach of AR is better and more effective compared to others, but it is a helpful complement to other technologies for managing a building life cycle.

Through AR it is possible to obtain a special feature, namely "instant visualization", which facilitates communication and decision making between the parties involved in the project. According to Wang J et al. [27], AR inherently involves human interaction with real and virtual information sources. Within BIM technology, the use of AR allows designers to place the virtual construction schematic in a real physical environment; it grants owners an engaging and interactive experience and suppliers the ability to communicate effectively with both clients and the technical team.

The case studies are different, the first being a road infrastructure project, more specifically a tunnel (already completed) and the second an architectural project for a commercial/service building (under construction), both with 3D models from BIM, which allowed visualization in AR. **Table 1** shows the SWOT analysis for use of augmented reality in BIM based on the above presented use-cases and key takeaways from those examples.

Future research should generally focus on three main themes: (1) improving the understanding of information transfer processes, (2) enhancing software solutions, and (3) better understanding the required information contained in BIM models.

By improving the understanding of information transfer processes, we mean determining the tasks that could benefit most from AR technology [28]. It would

Strengths	Weaknesses
• View and evaluate the object; • Transport of information; • Scale the object in the application; • Consult real-time information from the common data environment and request revisions; • Simultaneous real-time interaction;	• Difficult to position the objects in the real scene; • Technical limitation of the mobile device causes unsatisfactory results; • Large data transfers; • Close integration with BIM software solutions;
Opportunities	**Threats**
• Easier knowledge acquisition and transfer; • Next generation of augmented reality based BIM modelers; • Education • Live remote interaction	• Relying on augmented reality for specific BIM object information; • 3D BIM model not available; • Information not available; • Information is not in the proper file; • Incorrect information

Table 1.
Augmented reality in BIM SWOT analysis.

also be interesting to explore an alternative information flow where building information models could be generated or updated in a real 3D space, on mobile devices, rather than on 2D computer screens in offices. It is expected that building information modeling software vendors will improve their products in this direction, which would also be based on research such as that conducted or reviewed in this paper.

Computer market leaders say augmented reality will change everything. But right now AR still has many challenges to overcome (improving occlusion, the process of creating 3D content and asset quality, connectivity, computing power of devices, miniaturization of hardware components ...) and if AR on mobile phones is the first step and most AR experience for a few years, AR on smart glasses will completely change the user experience. With smart glasses, users will have access to hands-free experiences, with content displayed right in front of their eyes and a full augmented view of the real world.

Acknowledgements

The presented research was supported by Erasmus Mundus Joint Master Degree (EMJMD) - BIM A+ European Master in Building Information Modeling (https://bimaplus.org). Their support is gratefully acknowledged.

Conflict of interest

The authors declare no conflict of interest.

Author details

Angela Cristina De Hugo Silva[1], Metod Gaber[2] and Matevž Dolenc[3*]

1 BIM A+, São Paulo, Brazil

2 Kolektor Koling, Logatec, Slovenia

3 Faculty of Civil and Geodetic Engineering, University of Ljubljana, Ljubljana, Slovenia

*Address all correspondence to: mdolenc@fgg.uni-lj.si

References

[1] BIMDictionary. Building Information Modelling (BIM). https://bimdictionary.com (Accessed: 2021-06-25).

[2] Eastman C, Teicholz P, Sacks R, Liston K, editors. BIM Handbook: A Guide to Building Information Modeling for Owners, Managers, Designers, Engineers and Contractors. 2nd ed. New Jersey: Wiley; 2011. 99 p. DOI:doi.org/10.5130/ajceb.v12i3.2749.

[3] D. Richard and C. Harty, "Implementing 'Site BIM': A case study of ICT innovation on large hospital project," Automation in Construction, 2013; 15-24. DOI:10.1016/j.autcon.2012.11.024.

[4] P. S. Duston and H. Shin, "Key Areas And Issues For Augmented Reality Applications On Construction Sites", in Mixed Reality in architecture Design and Construction, Netherlands, Springer, 2009, 157 - 170. DOI: 10.1007/978-1-4020-9088-2_10.

[5] Meža S, Turk Ž, Dolenc M. Mint: Measuring the Potential of Augmented Reality in Civil Engineering. Advances in Engineering Software. 2015; 90: 1-10. DOI: 10.1016/j.advengsoft.2015.06.005.

[6] DALUX. https://www.dalux.com/gb/ (Accessed: 2021-06-26).

[7] Björk B-C, Penttilä H. Mint: A scenario for the development and implementation of a building product model standard. Advances in Engineering Software. 1989; 11: 176-187. DOI: 10.1016/0141-1195(89)90049-1.

[8] Russell P, Elger D. Mint: The Meaning of BIM. In: Proceedings of the 26th Education and research in Computer Aided Archtectural Design in Europe; 17-20 September 2008; Antwerpen: eCAADe; 2008. p. 531-536. DOI: 10.1515/zaw-2014-0022.

[9] Zhang J, Yu F, Li D, Hu Z. Mint: Development and Implementation of an Industry Foundation Classes-Based Graphic Information Model for Virtual Construction. Computer-Aided Civil and Infrastructure Engineering. 2014; 29: 60-74. DOI: 10.1111/j.1467-8667.2012.00800.

[10] Renzi, F.V. BIM Application in Construction Management[thesis]. Barcelona: UPC Universitat Politècnica de Catalunya; 2018.

[11] Saar C. C, Klufallah M, Kuppusamy S, Yusof A, Shien L. C. Mint: Bim Integration in Augmented Reality Model. International Journal of Technology. 2019; 10:7: 611-619. DOI: 10.14716/ijtech.v10i7.3278.

[12] Wolfartsberger J. 2019. Mint: Analyzing the potential of Virtual Reality for engineering design review. Automation in Construction. 2018; 104: 27-37. DOI:10.1016/j.autcon.2019.03.018.

[13] Milgram Milgram, Paul, and Herman Colquhoun. Mint: A Taxonomy of Real and Virtual World Display Integration. Mixed Reality, 1999. 5-30. DOI:10.1007/978-3-642-87512-0_1.

[14] Azuma R. T. Mint: A Survey of Augmented Reality. Presence: Teleoperators and Virtual Environments. 1997; 6:4: 355-385. https://www.ronaldazuma.com/papers/ARpresence.pdf. (Accesed: 2021-06-25)

[15] Kalkofen D, Mendez E, Schmalstieg D. Mint: Comprehensible visualization for augmented reality. Visualization and Computer Graphics, IEEE Transactions. 2009; 15:2: 193-204. DOI:10.1109/TVCG.2008.96.

[16] Li X, Yi W, Chi H-L, Wang X, Chan A.P.C. Mint: A critical review of virtual and augmented reality (VR/AR) applications in construction safety.

Automation in Construction. 2018; 86: 150-162. DOI:10.1016/j.autcon. 2017.11.003.

[17] Kivrak, S., G. Arslan, A. Akgun, and V. Arslan. Augmented Reality System Applications in Construction Project Activities.In: Proceedings of the 30th International Symposium on Automation and Robotics in Construction and tradMining, Held in Conjunction with the 23rd World Mining Congress; June 2013; Montréal: ISARC; 2013. p. 1560-1571. DOI: 10.22260/isarc2013/0175.

[18] Bajura M and Neumann U. Dynamic Registration Correction in Augmented-Reality Systems. In: Proceedings of Virtual Reality Annual International Symposium; September; North Carolina; 1995. p. 52-60. DOI:10.1109/vrais.1995.512495.

[19] Williams G, Gheisari M, Asce A.M., Chen P-J, Irizarry J, Asce M. Mint: BIM2MAR: An Efficient BIM Translation to Mobile Augmented Reality Applications. Journal of Management in Engineering. 2015; 31:1: 1-8. DOI:10.1061/(ASCE) ME.1943-5479.0000315.

[20] Meža S, Turk Ž, Dolenc M. Mint: Component based engineering of a mobile BIM-based augmented reality system. Automation in Construction. 2014; 42: 1-12. DOI: 10.1016/j. autcon.2014.02.011.

[21] D. H. Shin and P. S. Dunston, "Identification of application areas for Augmented Reality in industrial construction based on technology suitability," Automation in Construction. 2008. 882-894. DOI:10.1016/j.autcon.2008.02.012.

[22] Fu M, Liu R. The Application of Virtual Reality and Augmented Reality in Dealing with Project Schedule Risks. In: Proceedings of Construction Research Congress; 2-4 April 2018; New Orleans; 2018. p. 429-438.

[23] C. Woodward and M. Hakkarainen, "Mobile mixed reality system for architectural and construction site visualization", in Augmented reality - Some Emergent Application Areas, A. Y. C. Nee, Ed., InTech, 2011,115-130. DOI:10.5772/26117.

[24] Muhammad A.A., Yitmen I, Alizadehsalehi S, Celik T. Mint: Adoption of Virtual Reality (VR) for Site Layout Optimization of Construction Projects. Teknik Dergi. 2020; 31:2: 9833-9850. DOI:10.18400/tekderg.423448.

[25] Ratajczak J, Riedl M, Matt D.T. Mint: BIM-based and AR Application Combined with Location-Based Management System for the Improvement of the Construction Performance. Buildings. 2019; 9:118: 1-17. DOI:10.3390/buildings9050118

[26] Chalhoub J, Ayer S.K., McCord K. H. Mint: Augmented Reaity to Enable Users to Identify Deviations for Model Reconciliation. Buildings. 2021. 11: 77: 1-18. DOI:10.3390/buildings11020077.

[27] Wang J, Wang X, Shou W, Xu B. Mint: Integrating BIM and Augmented Reality for Interactive Architectural Visualisation. Construction Innovation. 2014; 14:4: 453-476. DOI:10.1108/CI-03-2014-0019.

[28] S. Dong, A. H. Behzadan, F. Chen and V. R. Kamat, Mint: Collaborative visualization of engineering processes using tabletop augmented reality. Advances in Engineering Software, 2013. 45-55. DOI:10.1016/j. advengsoft.2012.09.001.

Chapter 9

Design Thinking for Computer-Aided Co-Design in Architecture and Urban Design

Shuva Chowdhury

Abstract

Bringing the designer's concept to the non-design expert's communicative level requires a significant understanding of the communication media. Primarily the design communication depends on the type of the tools used. Virtual tools with their pre-set operability limit the designer's ways of interaction with the artefacts. This article proposes a framework for designers to interact with non-design experts through an enhanced communicative media. The design framework indicates steps of design thinking to develop the interface by understanding both the virtual artefacts' perceptual affordance to the users and the design task. The paper discusses about projects tested in three different scenarios, urban design, architecture, and product design. It concludes with the arguments on designers' role as authors of the system design.

Keywords: design thinking, design participation, laypeople, interface, virtual artefacts

1. Introduction

Since the development of technology, the participation of end-users used to be an essential part in the design thinking phases. Design thinking with the users leads to deal complex aspects of design context, either in architecture or urban design, with the active participation of the stakeholders. The ever-changing development of virtual instruments and their current state of communicative power open new possibilities to engage non-expert stakeholders, primarily citizens, in the design decision making stages. However, mostly, the virtual instruments come with their pre-set operability, limiting the designerly ways of interactions with the virtual artefacts. Developing the design engagement instrument by understanding the need and interest of the end-users can bring conclusive results from the participation. Such instrument development phases follow the primary concept of design thinking. The process usually a journey of trial and errors. It is also necessary to identify goals and objectives of the design participation and ensure participants' feeling of control in the decision-making process. Thus, this study proposes a framework for designers like architects or urban designers to develop instruments by understanding the scope of interactions with the virtual artefacts and their perceptual representation in the virtual environment.

In the beginning, the article briefly discusses the current trend of shift in participatory design, the concept of tools, techniques and artefacts in design processes and the concept of design thinking in the instrument design. The brief literature review helps to understand the underlining concept of the proposed framework that had been tested with three interfaces. The article argues that the current trend in the computer-aided participatory design process requires developing the instruments with an empathised mindset on possible users' interactions with the interfaces.

2. A shift in participatory design

Participatory design, a broad term, generally means design processes whereby expert designers and people not trained in design work together can engage in collective creativity [1]. It is a collection of design approaches, methods, tools, and techniques shared between many disciplines [2]. Over the past decades, a significant shift has occurred in participatory design practice, where the end-users are involved in the decision-making process. Current covid scenario has also pushed that trend to find rapid, inclusive design solutions for safeguarding citizens' healthy well-being. To tackle this critical time, design professionals have accepted information technology-based intelligent and inclusive design techniques as mainstream line methods for design communication. Designers now design the instruments for end-users' participation to make the process legit and inclusive. In cases, they take advantages of computational tools to include users' feedback during the initial stage of the design phases. To do that, the designers need to identify the design problem and have empathy on the nature of possible interactions that can be obtained with the instruments. These steps are crucial and require experts' eye to identify the design problem and empathise with the users' likely behaviour while interacting with the instruments.

Participatory design process has seen its maximum success in community design practice. The result of such creative engagement depends on careful orchestration of the process keeping in mind the need and interest of the community. This design approach enables citizens to provide meaningful input into the projects through sharing diverse opinions and aspirations. However, to come into conclusive results in architecture and urban design, the participatory modes become pseudo, as the design facilitators cultivate the impression of openness of the participants but retain decision-making in their hand. Arguments are there to utilise such a variety of participatory design approaches where the design performance depends on the system, as they are mission-oriented, and the knowledge is reduced to its instrumental value. But the current disruption in the democratisation of technology shows us that instrumental values can meet social values. In that regard, the mission-oriented participatory design approach can produce design decisions on the demand of citizen needs.

Participatory design practices, which primarily focus on designing the technologies and services, are mainly informed by the user-oriented design approach. According to Bratteteig et al. [3] an approach can have six phases as an iterative design cycle (**Figure 1**). In the cycle, both the process and product are of equal importance. The process defines the usages, and the artefact (product or service) allows those usages to explore in different stages [4]. The importance of this process lies in the technology which mediates the users' behaviour. Involving future users as co-designers significantly increases the chance that the design outcomes have values and meanings to them. However, when there is a discrepancy between the design context and usage context, the process will not be strong enough to mediate the behaviour of the users in the way it was envisioned.

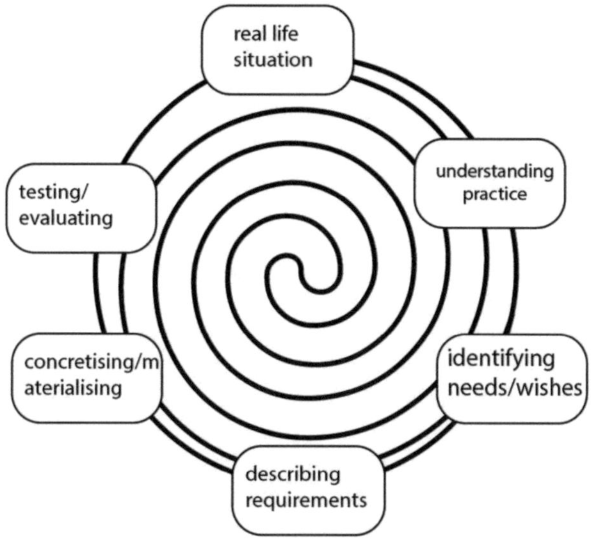

Figure 1.
Use-oriented design cycle [4].

3. Tools, techniques and artefacts

Participatory design instruments, including digital and physical games, mapping instruments, and visualisation systems, are used in participatory design activities to engage stakeholders in design thinking [5]. Participatory instruments support collaborative design enquiry and bring together a network of actors with different backgrounds, competencies and experiences [6]. Ehn [7] describes a design philosophy called 'tool perspective', which suggests that computer-based tools should be designed by considering the traditional practical understanding of tools and materials used within a given craft of profession. That means the design process must be carried by the common efforts of skilled, experienced users and design professionals. The users possess the needed practical understanding of the design but lack insight into the technical possibilities. The instrument designers must understand the specific decision process that uses the tool.

Traditionally drawings, product samples, models, and now virtual 3D models are used to mediate the journey of a building concept to the actual built form. These artefacts are often produced for different purposes and obviously for people with varying understanding of the design and construction process [8]. Luck [9] argues that design practice using physical artefacts at the early stages of building design is appropriate for design conversation, as it develops users' understanding of the design. The design conversation builds the user's confidence in the appearance of the design, rather than only through the ability of the artefacts to represent a future reality. The artefacts embody the current knowledge of the design in its present status and, during a conversation, prompt discussion on ideas. Here, the 'act of interpretation' is acting as a part of the design process. In this regards, Bucciarelli [10] stated earlier that design only exists in the collective sense and is realised through conversation and action. In the same note, Luck and McDonnell [11] observe that an architect has a range of prompts and conversational repertoires to elicit information from users when discussing a design.

The studies below primarily talk about the generation of virtual three-dimensional (3D) artefacts as elements for design conversation. The current nature of technological interfaces enables anybody to manipulate virtual 3D artefacts by simple hand device movement in virtual environment (VE). The tested instruments below can produce instantaneous 3D artefacts with perceptual understanding particular to the instrument and the design tasks.

4. Design thinking in instrument design

Computer-based instrument design requires a design thinking mindset. Generally, design thinking is defined as an analytic and creative process to engage users in opportunities to experiment, create and prototype models, gather feedback, and redesign [12]. According to the Interaction Design Foundation group, there are five-stage in the design thinking model [13]. They are as follows: empathise, define (the problem), ideate, prototype, and test (**Figure 2**). The first stage is to gain an empathic understanding of the design problem. It includes consulting the experts to collect information more on the concerning issue through observing and engaging with the users. In the defining stage, the designer put together the gathered information. This is where the designer will analyse the observations and synthesise them to define the core problems. In this stage, the designer has to define the problem in a human-centred manner. In the third stage, the designer starts to generate ideas, as the users' initial understanding and need have been analysed in the earlier stages. A new innovative solution can be attained through "think outside the box" strategies with a solid background of expert designers. Then, the steps of prototyping, where the designer test the developed instrument by himself or within a team. This is an experimental phase, and the aim is to identify the best possible solution for each of the problems identified during the first three stages. This stage helps get a better idea of the constraints inherent with the instrument and gives a clever view of how real users would behave, think, and feel when interacting with the instrument. Finally, the designer rigorously tests the complete product using the best solutions identified during the prototyping phase. It is an iterative process. The results regenerated during the testing phases are often used to redefine the problems and inform the understanding of the users.

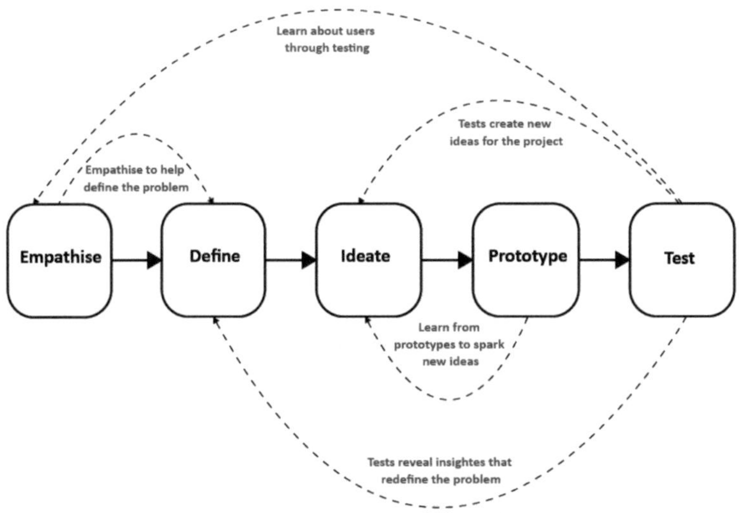

Figure 2.
Five-steps of design thinking [13].

5. The proposed framework

The proposed framework combines steps of initial data collection through a survey or other resources, instrument developments, design engagements and data collection (**Figure 3**). The initial step of the framework depends on the survey of the users' needs. It can be achieved through a questionnaire survey or by counting the brief of the clients. The second step is the instrument development steps. It comprises identifying the virtual instrument, defining the task by understanding the affordance of the instruments, and then finalising the design engagement interface. The instrument development stage is a non-linear loop-based iterative process. The instrument development steps itself is a reflective process where the design decision on instrument development depends on testing and self-evaluating approaches. Finally, the design engagement step, where the designer has to design the experiment setup to involve the non-experts in the design process and collect data from the engagement.

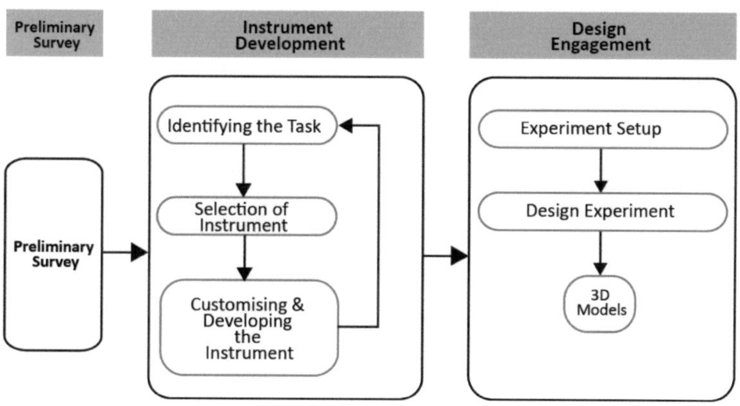

Figure 3.
The framework.

6. Interfaces for urban design

Here, two different scenarios have presented the immersive quality and the nature of communication. The instruments possess their specific quality of perceptual understandings of the artefacts in the virtual environment and, they also offer different ways of generating those artefacts. These differences instigate the needs to develop the design tasks along with the possible interaction with the users.

6.1 Immersive instrument

In 2019, an immersive virtual instrument developed to involve community people in the early stage of designing their neighbourhood [2, 14]. In the beginning, an initial preliminary survey conducted to identify the interest of the people. The research looked at a low-density suburb based in Wellington, New Zealand. The study looked at the local city council's charrette report to understand the design problem. The survey also helped to develop the design engagement tasks. These steps helped to understand the design engagement context. Then, in the instrument development stage, the designer had to select the right tool by empathising with users' possible interactions. This step was iterative. Several virtual software with different perceptual understandings 2D and 3D investigated to find a flexible tool for intuitive design generation of virtual artefacts. An immersive virtual instrument had been chosen, where the virtual 3D artefacts generated with the movement of the hand devices. A design collaboration set-up developed for participants to design as a team (**Figure 4**). Finally, a protocol analysis conducted to investigate the type of design communication [15]. **Figure 5** shows some moments of the immersive environment of the users and the participants from the community. The result showed that the employed immersive iterative 3D artefacts provide predictive

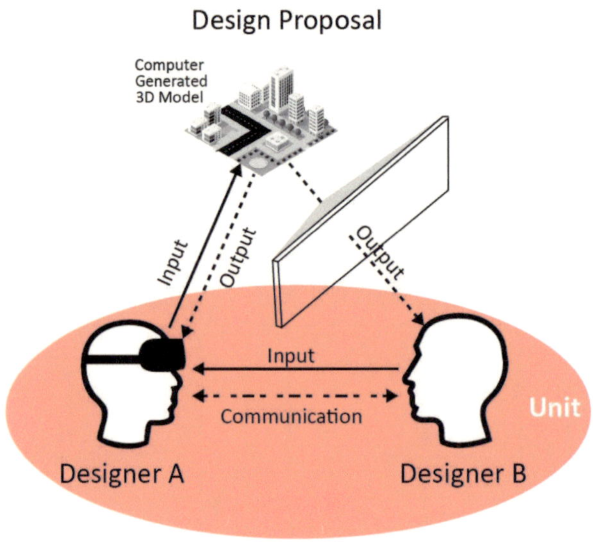

Figure 4.
Design collaboration unit.

Figure 5.
Laypeople's participation in virtual co-urban design.

and explanatory power for understanding the interaction. The design discussion progressed when every designer's action produced visual information and initiated the next level of design action. It happened due to successful design communication media, which provided continuous visual feedback to the co-designer. The design task and the set-up of the experiment oriented designers to create building forms. Designers mutually construct conversation moment by moment as a form of interactivity through their use of verbal exchange and design action. The employed virtual environment (VE) design process reflected design actions and negotiations between designers. VE design communication happened due to the presence of design interactions. It resulted from human-computer interactions, where the computer produced 3D artefacts in the VE and eventually provided visual feedback to take design actions and initiated design discussion among designers.

If we see the instrument development stages, it can be seen that the success of the research depends on the type of software selected and the way design tasks was dealt. To do that, the designers had to empathise the possible ways of users' interaction. This step requires frequent testing with all other possible instruments. Developing the design task is another challenging part of such creative engagement. The task should have relevance to the design problem. The virtual representation of the urban environment also had the quality to deliver the resemblance of the 3D artefacts with the actual urban setting. Thus, the participants felt that they designed for their future neighbourhood.

6.2 Non-immersive instrument

In 2018, a non-immersive 2D interface was developed to generate the urban form based on parameters related to building height, plot ratio, land division, construction cost, and building width [16]. It scripted in 'Grasshopper3D', which is a visual programming platform in Rhino. The developed script relied on a set of rules and instructions that could generate the desired outcome. The interface offered various site-specific variables which could be modified in terms of number in the sliders (**Figure 6**). In one stage, the script had been extended for online collaboration. The development of the instrument started with understanding the design context. Same as the immersive one, it dealt with the same neighbourhood in Wellington. The initial survey helped to understand the design problem. Then, we had to identify the urban parameters translated into the visual programming platform in the instrument development stage. In this case, the parameters in the sliders were building height, plot ratio, land division, construction cost, and building width. The users could play with those parameters and could visual the generated urban forms in the interface. While developing the instrument, several self-evaluation steps explored. These steps required the empathising attitudes on possible users' interaction. The study did not get enough data to report on the users' performance. However, it shows a viable way to involve design experts to develop the computational interface for non-experts to be a part of the design generation.

There is some limitation of this kind of interface. Due to the non-immersive interaction, the design tasks are basically controlled by the given parameter of the interface. The users had to decide on the generated urban forms without knowing how one parameter affected the others. Mostly, the users choose by playing with the numbers in the sliders. As the changes happen by the system, it is hard to affiliate with the new scenario by the decision taken on the parameters. These lose the connection of design decisions. Besides, the relationship between the parameters building height, ratio, cost and land division controls the design outcome, which lacks the intuitive nature of design generation. However, this interface allows taking decision in urban planning scale, which is ineffective in the immersive instrument.

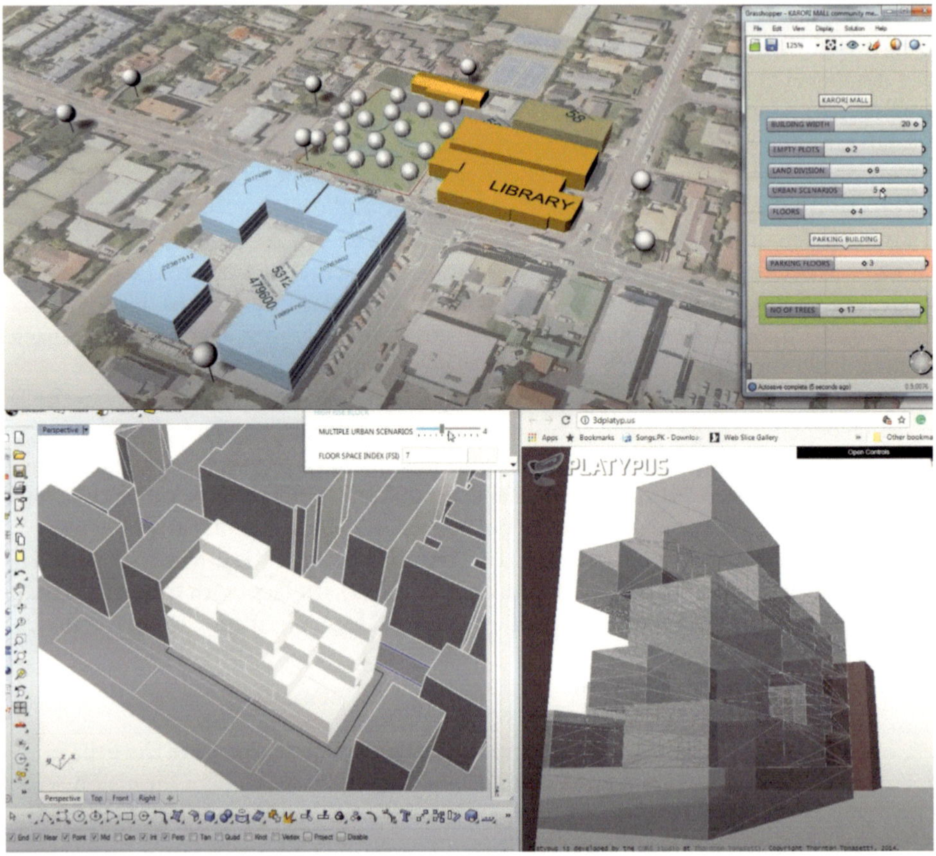

Figure 6.
Internet based 2D interface for urban form generation and collaboration.

7. Interface for product design

The interface developed for the product design is an online platform to generate parametric furniture. The interface developed in 'Grasshopper 3D' with its web-based extension 'Shapediver'. Similar with previous cases, defining the parameters for design interaction was one of the main tasks for the designer. The designer had to identify the parameters that could potentially modify geometry in relation to material cost (**Figure 7**). The users took their design decision based on negotiation between price and design. The instrument development started with empathising with the interface. As a part of testing, the survey conducted with seven participants. The task had been given to produce furniture which would cost less than 500 NZD. The online interface sent to the participants, and they generated the design by playing with the parameters and sent the model to the designer through the 'send the model' button. The users developed design options that were different to each other (**Figure 8**). The design variations show that the interface and its technological affordance can let first time users produce geometry on their choices. Though the perceptual understanding of the geometry is not in an immersive environment, the interface enables them to take relevant design decisions on the outcomes.

The study also explored in the remote collaboration environment. Designers continued their discussion through a text-based chat protocol. Designer A used

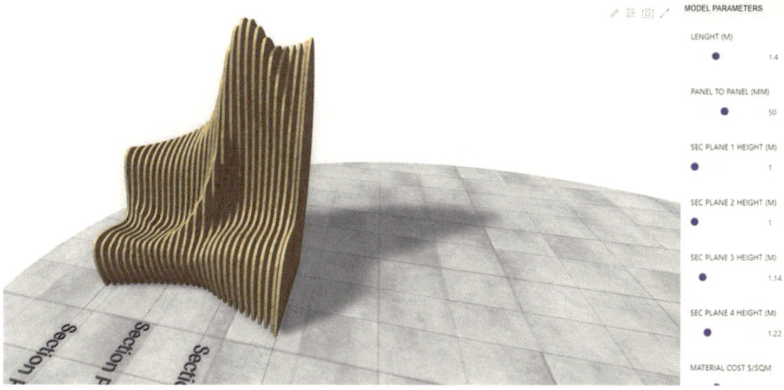

Figure 7.
The parametric furniture design interface.

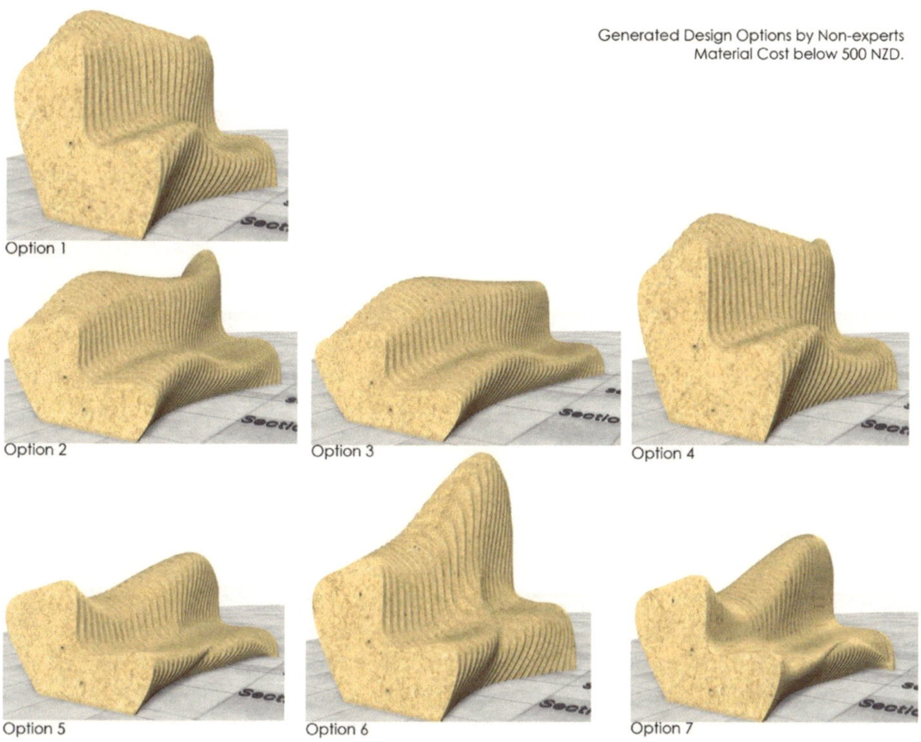

Figure 8.
The generated design options.

the desktop-based 2D interface to generate the furniture while designer B saw the instant outcomes through a screen share and provided responses in a chatting system (**Figure 9**). The collaboration aspect in this scenario is not instantaneous but working in an internet-based communication protocol. The users who have access to design commands need to familiarise with the interface and its opportunities to take design decision following the suggestion from the other end users. The co-designers are not getting control over their decision as one person decides on the other side of the network. There is a lack of affiliation exist in the system. However, the design set-up seems to be efficient in any remote settings.

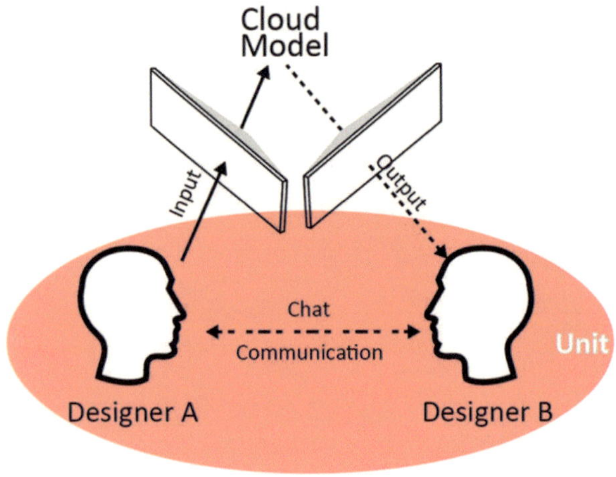

Figure 9.
The remote collaboration design unit.

8. Discussion and conclusion

Developing an instrument for end-users design interaction and participation requires experts' contribution to identifying the design problem, developing the task, and the instrument. These are evident in any design thinking approach. Experts play a significant role in determining the problem by empathising with evidence-based input through a survey or other recorded resources. They also need to synchronise the design task with the design problem of the context and the instrument. At the beginning of selecting the computational instrument, the experts rely on self-evaluative techniques to empathise with users' possible interaction. In cases, they test with participants to assess the system. Above, the three examples have not informed about those initial testing elaborately; however, the framework indicates the scope of possible ways of initial pre-testing before inviting the non-experts in the creative engagement. Besides, to develop any computer-aided design instrument, the experts also need to know how the computer system can deliver an interface for first-time users. Usually, finding the design problem is the designers' (both architects and urban designers) tasks; however, to deliver a computer-aided instrument, they need to understand the working process of a computer system. All the investigations show that, in this digital era, designers now need to understand the computer tools and their scope of communication to involve non-experts in the design process. The studies also demonstrate how virtual artefacts can influence users' perceptual understanding in the VE. Generally, the nature of design decisions depends on the interface types and their quality of visual communication with the users. The users decide on their design action against the instant visual feedback they get in the VE. If the interface is a 2D desktop, in that case, the interaction happens in a non-immersive environment, where the perceptual understanding of the 3D artefacts remains as scalar objects of the representation. In the immersive virtual environment, the users get a perceptual experience of the design content in full-scale. That means, whether the interface 2D or 3D, it is evident that interfaces offer easy manoeuvring systems to the non-experts. In addition, the collaborative design setup allows them to act as a design team to take decisions as co-designers. In conclusion, for all cases, the design thinking framework follows the steps of a defining problem with a participatory mindset, selecting a tool with relevance to the design task, testing the instrument and designing the engagement pattern. These are the principles for any design thinking research, where the experts with computer-aided design generation in architecture and urban design would make the process inclusive and informed.

Disrupted digital technology is already changing the way architecture is designed and built. Now, the architecture of industrialization has replaced by the architecture of information. These profound changes are primarily centred on the foundations of architectural practices, such as representation, information management, and virtuality. Besides, the revival of social reformation encourages the inclusion of participatory design approach in the architectural design process. To continue a design process without creating repetition and to avoid homogeneity and recurrence of the same, one needs to reinvent the design continuity through the spontaneous participation of people. It means design iteration can be spontaneous if people are engaged in the design process. The recent advancement in the computing power of digital technology supports the continuity of design iteration, which possesses representation techniques for non-experts to involve in the design process. Such technology-driven design representation and iteration tools allow non-experts to participate actively in design ideation and generation stages. It reduces the power gap between non-experts and experts and changes the role of experts from designers to system designers.

Author details

Shuva Chowdhury
Southern Institute of Technology, Invercargill, New Zealand

*Address all correspondence to: shuva.chowdhury@sit.ac.nz

References

[1] Sanoff, H. (2011). Multiple views of participatory design. *focus, 8*(1), 7.

[2] Chowdhury, S. (2020). Virtual Environments as Medium for Laypeople's Communication and Collaboration in Urban Design.

[3] Bratteteig, T., Bødker, K., Dittrich, Y., Mogensen, P. H., and Simonsen, J. (2012). Methods: Organising principles and general guidelines for participatory design projects *Routledge international handbook of participatory design* (pp. 137-164): Routledge.

[4] Van der Velden, M., Mörtberg, C., Van den Hoven, J., Vermaas, P., and Van de Poel, I. (2014). Participatory design and design for values. *Development, 11*(3), 215-236.

[5] Sanders, E. B. N., Brandt, E., & Binder, T. (2010). A framework for organizing the tools and techniques of participatory design. In Proceedings of the 11th biennial participatory design conference (pp. 195-198).

[6] Brandt, E., Binder, T., & Sanders, E. B.-N. (2012). Tools and techniques: ways to engage telling, making and enacting Routledge international handbook of participatory design (pp. 165-201): Routledge.

[7] Ehn, P. (2017). Scandinavian design: On participation and skill *Participatory design* (pp. 41-77): CRC Press.

[8] Lawson, B. (2012). *What designers know*: Routledge.

[9] Luck, R. (2007). Using artefacts to mediate understanding in design conversations. *Building Research and Information, 35*(1), 28-41.

[10] Bucciarelli, L. L. (1988). An ethnographic perspective on engineering design. *Design Studies, 9*(3), 159-168.

[11] Luck, R., and McDonnell, J. (2006). Architect and user interaction: the spoken representation of form and functional meaning in early design conversations. *Design Studies, 27*(2), 141-166.

[12] Razzouk, R., and Shute, V. (2012). What is design thinking and why is it important? *Review of educational research, 82*(3), 330-348.

[13] Dam, R. F., and Siang, T. Y. (2021). 5 Stages in the Design Thinking Process. Retrieved 23 May, 2021, from https://www.interaction-design.org/literature/article/5-stages-in-the-design-thinking-process

[14] Chowdhury, S., and Schnabel, M. (2020). Virtual environments as medium for laypeople to communicate and collaborate in urban design. *Architectural Science Review*, 1-14.

[15] Chowdhury, S., and Schnabel, M. A. (2019). Laypeople's Collaborative Immersive Virtual Reality Design Discourse in Neighborhood Design. *Frontiers in Robotics and AI, 6*(97).

[16] Chowdhury, S., and Schnabel, M. A. (2018). *An Algorithmic Methodology to Predict Urban Form.* Paper presented at the Proceedings of the 23rd International Conference on Computer Aided Architectural Design Research in Asia, Beijing.

Chapter 10

Plenum a la Mode - Augmented Reality Fashions

Geoffreyjen Edwards, Jonathan Caron-Roberge, Dominique Michaud and Jonathan Proulx Guimond

Abstract

Inspired by ideas portrayed in science fiction, the authors sought to develop a set of augmented reality fashions that showcased scenes from a science fiction novel recently published by the principal author. The development team included artists and designers, a programmer, and the writer. Significant technical challenges needed to be overcome for success, including fabric construction and manipulation, image enhancement, robust image recognition and tracking capabilities, and the management of lighting and suitable backgrounds. Viewing geometries were also a non-trivial problem. The final solution permitted acceptable but not perfect real-time tracking of the fashion models and the visualization of both static and dynamic 3D elements overlaid onto the physical garments.

Keywords: augmented reality, fashion, belts, tracking, visualization, weaving, integration

1. Introduction

As a writer of science fiction, one of us (Edwards) has written about the use of dynamically changing skin-tight garments in a future civilization. In the scenario worked out in this fictional world, the first volume of which is called Plenum [1], these membrane-like garments protect their wearers from the hardships of the vacuum of space, but they also incorporate nanotechnologies that allow surface patterns to change and swirl across the surface of the skin, or even to extrude into the third dimension.

Inspired by these ideas, and also by earlier work as a fashion designer on the part of Edwards (see collections at www.geoffreyjenedwards.com/fashion-design), a small multidisciplinary group was pulled together from different backgrounds and abilities to prototype garments that do this. Rather than using nanotechnologies, which aren't advanced enough to permit this kind of application yet, we are working with augmented reality (AR) technologies instead. Michaud is a student with training in mathematics who has retrained in the textile arts, Proulx Guimond is a visual artist with training in 3D design and animation, and Caron-Roberge is a programmer.

Augmented reality is beginning to be used by certain avant-garde fashion designers and artistic visionaries [2]. Like many emerging technologies, the technical challenges are considerable. The field of augmented reality capabilities has progressed rapidly in the past decade, but there are still major issues that are not easily resolved to support real-time AR visualization in a robust artistic performance setting.

In this paper, we shall describe the application context, the particular choices made, and the technical challenges we overcame in order to present a real-time fashion show in front of a live audience, showcasing our designs.

2. Augmented reality technology

Augmented reality is achieved by providing elements in a scene which can be recognized by computer vision software and tracked over time [3]. Once recognition has been obtained, the coordinates associated with the recognized image, including its orientation in space, then become the anchor point for overlaying 3D virtual reality elements in relation to this anchor point. Often AR involves careful design to ensure that the virtual elements merge seamlessly into the real elements from the observer's perspective, so that it becomes difficult to distinguish what is real from what is virtual. That is the magic of AR, that it provides an enhanced or enriched image of the real world.

The task of creating computer vision software to do the recognition and tracking is a formidable one. Fortunately, a number of APIs now exist that do a lot of the basic work, and instead the programmer can focus more on the interface design and visual integration, which is also a challenging problem. We selected the Vuforia™ API after studying several alternatives [4]. This API is sufficiently mature to support robust work in AR. On the downside, however, it offers few possibilities for adjusting the recognition parameters. In a sense it must be treated as a black box.

In earlier work, we used artificial high contrast patterned targets, such as the one shown in **Figure 1** [5]. Indeed, following numerous studies and trials in the early years of work on AR, these patterned targets had become the de facto standard for ensuring reliable recognition and tracking [3]. However, although these patterns

Figure 1.
Example of our initial effort using a printed pattern target.

still provide the best targets, the APIs now support non-standard targets - in fact anything that offers high contrast in an asymmetrical pattern can act as a target. Furthermore, Vuforia offers a cylindrical recognition mode. Essentially, instead of recognizing a flat target, the software will recognize a cylindrical target, which can be tracked even as it turns, thus allowing virtual elements to be added in relation to the cylinder which follow as it turns. For applications involving garments, this is ideal.

We use the Unity Game Engine™ as our programming or interface environment. Unity provides a stable, robust, integrative framework for immersive reality projects, and interfaces directly with Vuforia [4]. We have used it in the past for both virtual reality [6, 7] and augmented reality applications [5].

3. The fashion event

As indicated earlier, we conceived of an event that would both showcase the fashion designs and also the science fiction books which inspired the creation of these garments. As a fashion designer with several collections, Edwards had experience mounting fashion shows. We therefore conceived of an event built around the idea of a fashion show, that is, with a runway, models, and successive reveals of different garments, combined with music, an audience, and lighting ambiance. The event, however, also sought to integrate elements usually associated with a book launch, that is, excerpts to be read, copies of the book available for purchase and a book signing activity. Finally, we also undertook to develop additional representations of the event via video, and an app that could showcase some of the designs developed. These could serve both to document the work achieved and also act as marketing tools for the event and for the books that inspired the event.

Edwards selected several excerpts from Plenum that both acted as examples of the writing but also highlighted settings which could be visualized. We therefore conceived of four scenes (initially three - one was added later in the development) that would be presented. For each scene, we designed a different belt using colors and textures that matched those of the scene. Some of these original design choices were modified however, over the course of the work, as a result of changing conditions external to the work itself (and in particular, constraints imposed by the Covid19 pandemic, which struck about six months into the work).

The three initial scenes selected were (a) the approach by the main character to an artificial habitat build around a small moon or moonlet, (b) the climactic final scene which takes place in the vicinity of a sun, and (c) the large spaceborne platform called the Annex within which the main character spends most of the time. The scenes also include two kinds of ship, artificial ships and organic ships called jonahs, which are descendants of Earth-bound sperm whales. The fourth scene eventually added was to showcase the kind of dynamically shifting skin textures described in the books. For this we selected a fractal pattern.

Here are examples of the first two excerpts [1]:

Excerpt #1: "The Rock, as its residents named it, was some fifty kilometers in diameter ... Naively, one would have expected vertical structures pinned onto the Rock, but the Yard was structured horizontally rather than vertically. Instead of spires, it consisted of folded sheets through which poked a large, rounded section of the Rock. It had slightly more gravity than a true microgravity environment, but at humorn tempo, objects still settled quickly enough to notice. If you left things floating, they drifted downwards at a stately pace, but the oblique angle of the sheets often made the downwards direction something other than the apparent vertical!"

Excerpt #2: "Vanu boarded the flyer, and piloted it across the short distance to the access port for the Wellhead. This was an egg-like structure which looked to be tethered to the star by a gossamer thread that dropped away beneath it. The construction looked ungainly, a beached whale thrown up by the seas it served, although Vanu had to dampen the brightness of hir visual field to even perceive it properly. Vanu also had a momentary thought about what the Wellhead represented. It was like a guardian, or a gatekeeper, a Saint Peter at the pearly gates, not just to support those who went down into the Core, but also to screen out the unworthy."

4. Fashion and textile choices

Although our end goal is to create full body AR fashions, for technical reasons this is still too ambitious an undertaking. Obtaining one set of AR elements to follow the garment is already a significant challenge. So we have focused on the creation of a set of belts. One of our reasons for choosing belts is a consequence of the technology we are using, that is, the cylindrical recognition feature of the Vuforia API. However, it took considerable experimentation to work out how to do this efficiently. For example, although we earlier used patterned targets printed on paper (**Figure 1**), we wanted to develop fabric-based targets, but fabrics have a stitch structure that acts to effectively degrade the contrast and resolution features of the target images.

Tests with different woven fabrics were undertaken first to determine which features are best for ensuring recognition by the Vuforia software. The image used must be non-repetitive and asymmetrical as well and must favor sharp linear edges rather than curves. Experimentation showed that high contrasts are required for the Vuforia recognition and tracking software to function effectively. Furthermore, recognizable scene elements must be evenly distributed across the cylinder to ensure the belt can be tracked adequately over time and body movement. Some occlusion, for example from the arms and hands, can be tolerated without losing the tracking lock. We initially tested printed images such as the one shown in **Figure 2**.

Although the Vuforia API supports a range of possible cylinders, for our application we needed a wide, relatively narrow cylinder, since it was to be worn as a belt and remain flat in its lengthwise dimension, thereby keeping its design stable. After some experiments varying the width of the belt, we found that it needed to be at least 15 cm wide to provide a reliable recognition lock. This is wider than most belts, but there are many examples in fashion where belts of that width are used. We also discovered that the Vuforia API was happier if the belt had a high contrast edge (see **Figure 3**).

Images can be pre-tested in Vuforia before they are definitively used. Essentially, Vuforia in its test mode can provide a map of recognized points (**Figure 4**,

Figure 2.
Example of a printed image used to test the cylinder recognition process.

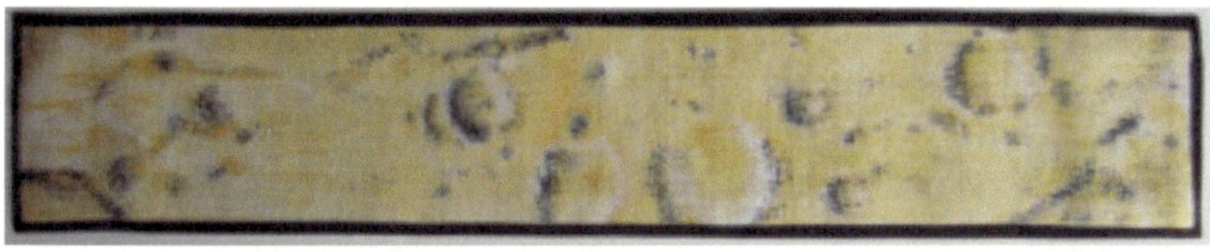

Figure 3.
The finished moon belt with its black edging.

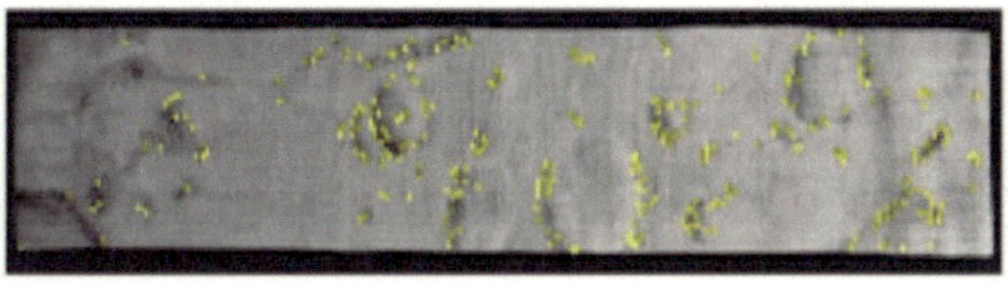

Figure 4.
Recognition results for the moon belt.

corresponding to a segment of the belt shown in **Figure 3**). Hence if the density of recognized points is too weak, the image can be further manipulated to enhance recognition, but only up to a point. Furthermore, in the case of a woven belt, such as shown in **Figure 3**, modification poses a challenge. Although we tested the scene recognition initially using printed images (e.g. **Figure 2**), the final tests had to be done with the finished weave, as the Vuforia software is sensitive to small deformations in the image. However, once the weave, itself a time- consuming step, is finished, further correction is difficult. Since some contrast is lost with the move to the woven image, we had to be creative to seek better recognition at that stage of development. We did attempt to further enhance the contrast using embroidery techniques, but ultimately the solution was of a different nature (see the discussion below concerning the background).

Weaving complex scenes such as the image of the lunar surface shown in **Figure 3** could be done accurately on a Jacquard loom. Although there are a number of Jacquard looms in Quebec City, access to them proved to be difficult (and would have been virtually impossible once the pandemic arrived). Fortunately, Michaud had completed a degree program in the textile arts and was able to use a range of different techniques to achieve similar results.

Considering the restriction of available equipment, Michaud used her own four-shaft jack table loom to test three fabric designs; overshot, shadow weave, and a double ikat-like technique (without resists). Cotton was used, as it is easy to find, does not stretch, is affordable, and can be dyed if necessary, with thread counts of 16/2, 8/2, and 8/4. Overshot in black and white, with the bigger thread 8/4, proved to be the most effective. However, the results were deemed too abstract for the project, so a compromise was made for a more pictural solution, using a double Ikat technique [8, 9], which offers a way of painting the threads before they are woven. Tests were undertaken to determine the appropriate weave density and to validate the painting techniques.

The second belt produced used an image of the solar surface (**Figure 5**). Recognition was generally stronger than it had been for the lunar surface (**Figure 6**). The belts for scenes 3 (space habitat) and 4 (fractal textures) are still under development. More details will be given below, since other parts of this work are in a more advanced stage of development.

Also, in order to ensure recognition, the belts needed to be reinforced to make them stiffer. Otherwise, they would bend to conform to body shape and thereby

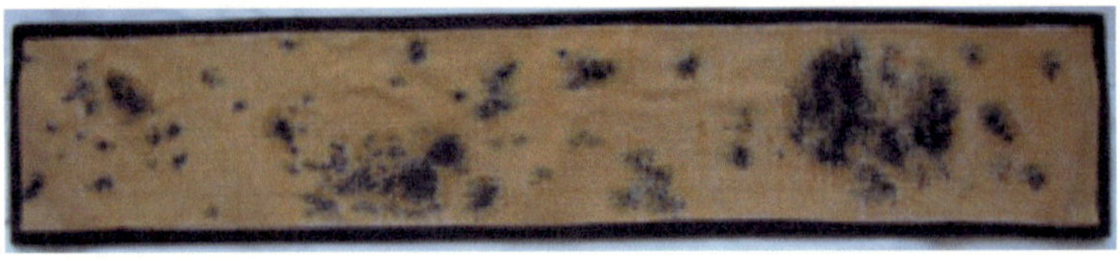

Figure 5.
The finished solar belt with its black edging.

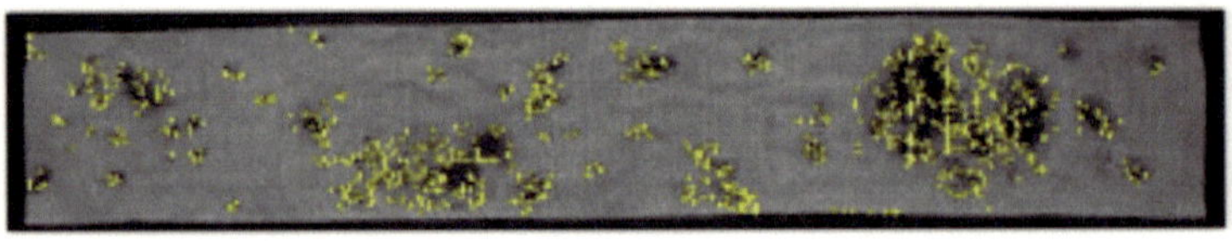

Figure 6.
Recognition results for the solar belt.

distort the image viewed by the Vuforia software, disrupting recognition. A canvas backing was therefore added to the inside of each finished belt. The belts are 32 in. in circumference, that is, they were designed for the slender physique of most fashion models, and an elastic closure was added so that they can be adjusted to different sized waists. Experimentation has shown that the presence of a small gap in the image at the back does not appear to disrupt the recognition and tracking, nor does a small overlap.

Each of the four scenes is characterized by a different color palette. This was, of course, intentional, to give a distinct character to each scene/belt combo. The lunar scene presents yellow and orange elements, while the solar scene is more red and orange. The platform itself is gray, but the nebula is predominantly blue and green with splashes of red, purple and yellow. For the fractal scene, we decided to produce the patterns in blue and white rather than black and white. This gives its corresponding belt a distinctive color.

5. 3D virtual elements

As illustrated in the excerpts given above, the scenes depicted are highly visual in nature. Proulx Guimond is a visual artist with whom Edwards had worked previously. Indeed, he developed the cover art for the novel, Plenum [1]. He created 3D scenes for each of the four situations. The scenes were designed so that they would integrate with the visual appearance of each belt. Hence the lunar scene was anchored to the belt which presents the surface of the moonlet itself (**Figure 7**), showing as extrusions of the artificial city constructed around the moonlet (in yellow). For each scene, we planned three distinct elements, first a static 2D visualization, second a static 3D visualization and finally a dynamically changing 3D visualization. The dynamic elements added to the depiction of the moonlet and its artificial extrusions were the arrival of starships which would dock with the city.

In the second scene, the heliocentric orbital station is modeled (in exaggerated scale, since it would normally be too small to see at the scale of the sun) along with the surface of the solar photosphere with its sunspots. Furthermore, the dynamic elements included the jonahs and a solar flare made for a dramatic and flamboyant event within the whole sequence. To provide the fashion representation with more drama, the orbital station was situated on a diagonal with respect to the solar surface.

Figure 7.
3D image constructed for the lunar scene.

Figure 8.
3D image constructed for the space platform scene.

In the third scene, the main focus was the spaceborne platform itself, again, oriented obliquely with respect to the model's body and extruding from this (**Figure 8**). The platform includes a dock and two rotating cylinders used to provide pseudo-gravity to its occupants. Here, again, jonahs are seen moving past the structure, along with an octopus who is also one of the main characters in the story. The platform is viewed against a colorful emission nebula in the background, and here the dramatic event is the swirling movement of the nebula under time acceleration, the spinning of the cylinders and the movements of the jonahs and octopus.

The 3D graphics were created in 3DS Max™ and eventually transferred to the Unity environment. Efforts were undertaken to make the render of the visual as light as possible to ensure real time updating of the scene elements. This was achieved by reducing the complexity of the scene to its bare minimum, that is, incorporating only a small number of simple elements, and limiting the number of different materials and textures used. In addition, the animated sequences were also kept to a minimal complexity.

The visual integration of the virtual scene elements with the physical belt and, indeed, the body of the models required further adjustments once all the elements were in place and fully integrated. The integration, as outlined below, posed significant challenges since the majority of the work was carried out individually under lockdown conditions over the course of 2020 and the first half of 2021, that is, during the midst of the Covid19 pandemic. Indeed, at one point we considered doing the fashion show as a purely virtual event. However, the initial impetus for the project was for a show with live audience, and we remained convinced that the full impact of the AR technology consists of its use in real time in the presence of a physical audience. We therefore decided to delay the event itself until it could be performed live. As a consequence, however, this paper is being completed before the final presentation event.

6. Garment tracking

Testing of our efforts to integrate the tracking software with the belts revealed a succession of challenges. We found that the software was temperamental in its ability to recognize the belt patterns, even though initial efforts had been successful. Good lighting turned out to be an important element in ensuring this recognition. Once recognition was achieved, and the virtual elements reliably overlaid on top of the optical image, it would retain the lock for a certain set of manipulations, but eventually the lock would be lost when viewing conditions became less than ideal (for example, if the person wearing the belt moved too far away from the camera). The model would then need to move in close, or image viewing parameters otherwise manipulated, until a recognition lock could once more be obtained. At first, the virtual image was stable only for short intervals. We introduced some persistence into the visualization, so that even when the recognition failed, the virtual objects would persist a second or so. This sometimes results, however, in a jerky movement of the virtual elements in relation to the model's movement.

Effective integration also required introducing occlusion effects. Hence, we used the belt's cylindrical shape to create an occlusion model and used this to hide virtual elements as they ostensibly moved behind the body. Without intentionally occluding the virtual elements in this way, they were perceived as always being in front of the model, breaking the illusion.

As the model moved, the location of the virtual elements would often lag behind the model. We used a cube-like envelope to test these issues. We eventually discovered that we had been testing the performance of the visualization against "busy" backgrounds. Much of the early testing was done in a living room or kitchen viewed on Zoom. When we met and did tests outside, the same problems persisted until we chose a uniform background. At that point, many of the tracking problems diminished. After that, we chose uniform walls for testing. Feature-filled backgrounds clearly confused the recognition software, causing it to lose its lock more frequently. Once this problem was identified, lock persistence, although not perfect, became acceptable.

At one point we also sought to use an accelerometer and Inertial Motion Unit (IMU) that could be integrated into the garment as another source of data to help stabilize the imaging. Our efforts in this direction failed to generate the necessary stability, however, and the effort was abandoned.

Another strategy we eventually developed to ensure longer lock persistence was to restructure the viewing geometry using a second camera feed. Essentially, we had

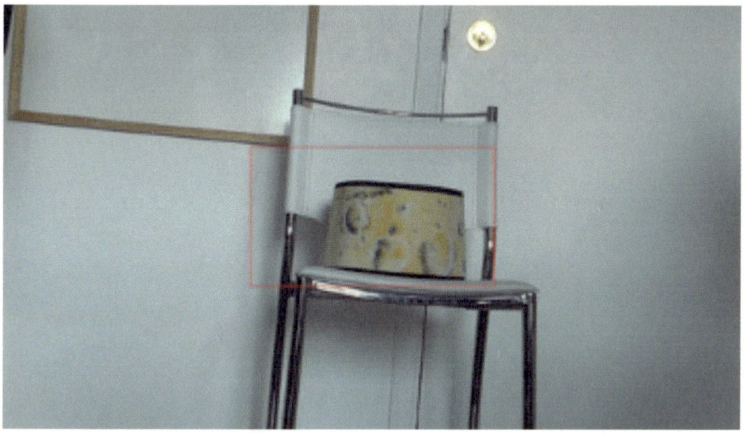

Figure 9.
Image of the finished lunar belt in situ along with image subset used for recognition.

been using the same camera image as the source for recognition and to serve as the support for the visualization. Our realization that Vuforia must degrade the image somewhat during processing, a realization that followed its lack of sensitivity to difference resolution webcams, led us to segment out of the image the region surrounding the belt and provide this image segment to Vuforia in a separate camera feed instead of the full image (**Figure 9**). This resulted in better recognition at a wider range of distances. Hence, for example, in our earlier attempts a distance of about a meter was required to ensure a recognition lock, but after the segmentation step was added, a reliable lock up to three meters could now be achieved.

Recourse to a more complex set of processing options such as described here meant, however, abandoning our commitment to permitting the use of a tablet or smartphone for viewing the AR scene. In the final configuration, we need access to a full computer, although a laptop is also acceptable.

7. Final integration (including music and staging)

Each belt is used to anchor its corresponding scene—indeed, each scene was designed so as to align the final appearance including the moving model with the esthetics of the 3D images. In principle, this means that several belts and animations could be viewed simultaneously. In practice, however, since each set of animations requires significant computing power to render in real time, we preferred to keep each presentation separate. Furthermore, the belts are worn with a white top to provide a canvas over which the 3D visuals can be overlaid.

The final integration of the 3D graphics with the physical belts are shown in **Figure 10** for the lunar belt and **Figure 11** for the solar belt. As indicated earlier, two additional scenes are under development, one involving a large space platform and the other a dynamically changing costume incorporating fractals. The 3D images are almost complete and the belt development is in its final stages. With the relaxation of lockdown restrictions currently underway, the organization of the event itself is also now possible.

A staging scenario was developed for the presentation in front of a live audience. This script included information on the timing of each segment. This was necessary also to assist in the development of the music accompaniment. At this point in the development of the event, the music has still not yet been finalized. However, we

Figure 10.
3D image for the lunar scene integrated with the physical belt in situ.

Figure 11.
3D image for the solar scene integrated with the physical belt in situ.

are proceeding as we have in previous projects, which is to mash together public domain samples of music into a coherent and dynamically interesting whole [6].

The total duration of the final show is estimated to be about 40 minutes. Following an introduction, the presentation of each of the four belts cycles through the three levels of visualization (static 2D, static 3D and dynamic 3D), along with the reading of the text excerpt, and hence each belt takes about ten minutes to present. After the individual belts are presented each in turn, the four belts are presented together in a kind of finale, without the full dynamics so as to render the processing tractable and with an additional excerpt. Following the conclusion, a book signing session is expected to be announced.

Our plan is also to stream the event for virtual participants. Indeed, the book publisher is located in Boulder, Colorado, and both collaborators and other interested parties are located there. Streaming became a de facto standard with the pandemic, and we retain that here. In addition, we are developing an app which will allow the 3d scenes to be viewed independently of the belts, as a value-added product to be made available to individuals as part of the marketing of the book.

8. Conclusion

In summary, our Plenum A La Mode event constitutes a unique amalgam of different ideas and technologies that straddles the artistic, scientific and technological worlds. The technical challenges in terms of obtaining reliable recognition of the fabric belt patterns, tracking the belt over a complex set of manipulations and movements in real time, and visually integrating both the optical scene with virtual three-dimensional enhancements were considerable. Furthermore, the project required the close collaboration of several individuals with different areas of expertise to achieve the desired results, including the writer, event and fashion designer (Edwards), 3D graphics artist (Proulx Guimond), textile artist (Michaud) and programmer (Caron Roberge). Another colleague with expertise in the development of music is also working with us, Jocelyne Kiss.

Among the challenges addressed and overcome were the introduction of occlusions using the belt cylinder as a model, so as to create the illusion of virtual elements passing behind the model's body, the provision of stable and good lighting conditions to ensure robust target recognition, the use of uniform backgrounds so as to limit confusion for the recognition software, the use of a second camera feed so that the tracking could be carried out independently of the scene enhancement, and the introduction of persistence to enhance the stability of the visualization of the virtual elements and prevent loss of the recognition lock.

The final result of these efforts provides acceptable, but not perfect, real time augmented reality image enhancements. The virtual elements experience a small amount of jitter while the software tries to track the belt, and if the movement is too rapid, the model passes behind a visual obstacle, or simply moves too far from the camera, the recognition lock is lost and the virtual elements disappear until a new lock is acquired (generally by moving towards the camera and holding still). The public staging requires the use of a desktop computer of relatively strong power, loud-speakers and a webcam. Our expectation is to project the enhanced image onto a screen as the model parades along the runway. The belts are themselves esthetically interesting as well as serving as target patterns for the augmentation.

Ultimately, we believe that more complete full body augmentations could be achieved using several targets integrated into the garment design. These might include flat targets in some areas, such as across the chest or back, as well as additional cylindrical targets for, example, around the thighs and arms. The augmented images would need to be matched to each other and integrated visually into a seamless whole, which will pose many additional technical challenges to the ones we encountered. Details of the final fashion show event will be made available through the principal author's website at www.geoffreyjenedwards.com.

Author details

Geoffreyjen Edwards*, Jonathan Caron-Roberge, Dominique Michaud and Jonathan Proulx Guimond
Centre interdisciplinaire de recherche en réadaptation et intégration sociale (CIRRIS), Université Laval, Québec, Québec, Canada

*Address all correspondence to: gedwoods2002@yahoo.it

References

[1] Edwards, G., Plenum: The First Book of Deo, Boulder: Untimely Books, 2021.

[2] Geroimenko, V. (ed.), Augmented Reality Art: From an Emerging Technology to a Novel Creative Medium, Berlin: Springer, 2014.

[3] Schmalstieg, D. and T. Höllerer, Augmented Reality: Principles and Practice, Boston: Addison-Wesley, 2016.

[4] Borycki, D., Programming for Mixed Reality with Windows 10, Unity, Vuforia, and UrhoSharp (Developer Reference), Microsoft Press, 2018.

[5] Edwards, G., Wearable Technology and Body Mapping, Chapter 9 in Applying Body Mapping in Research: An Arts-Based Method, K.M. Boydell (ed.), New York: Routledge, 92-103, 2021.

[6] Edwards, G., J. Kiss, E. Morales, C. McLaren, S. Lacasse, J.N. Falcon, J. Proulx Guimond and M.L. Bourbeau, Designing an Interactive and Participatory Opera, in Interactive Multimedia, D. Cvetkovic (ed.), Intech Publications, 2018.

[7] Arfaoui, A., G. Edwards, E. Morales, and P. Fougeyrollas, Designing Interactive and Immersive Multimodal Installations for People With Disability, in Virtual Reality and its Application in Education, D. Cvetkovic (ed.), InTech Publications, 2019.

[8] Harris, J. (ed.), 5000 years of textiles, Washington DC: Smithsonian Books, 2010.

[9] URL: https://prochemicalanddye.net/downloads/dl/file/id/115/product/0/warp_painting_on_cotton_silk_directions.pdf

Chapter 11

Augmented Reality as an Emerging Technology to Promote Products and Services

Alex Pacheco, Kevin Sánchez, Rosario Pariona, Nestor Cuba and Bertha Larico

Abstract

Augmented reality is a technology that provides more interactive advertising, where you can manipulate and be part of it with greater clarity and empathy as possible. In times of the digital era, companies in the retail sector face a new consumer profile, more digital, more aware and more informed, so companies are in constant competition to impress their customers. That is why this research seeks to describe the importance of augmented reality to promote products and services as an innovative alternative that captures attention and influences the customer's purchase decision. The research is of an experimental, quantitative and quasi-experimental design with a non-probabilistic sample whose evaluation instrument was a self-elaborated questionnaire that was applied to 60 randomly chosen Marketing managers from the retail sector. Obtaining as a result that augmented reality improves the level of engagement with the brand, generates greater added value and improves the level of innovation in advertising campaigns. What contributes to generate a different and innovative experience to advertise products and services through an emotional connection with people through mobile devices.

Keywords: augmented reality, benefits, promote, products, services, emergent technology

1. Introduction

Augmented reality is an emerging technology that allows improving the sale of products and services through the use of applications on mobile devices [1], highlighting the ease, usefulness and compatibility of use [2]. It helps to position itself in the minds of consumers, generating a differentiating factor with greater acceptance over competitors [3]. Likewise, Pizarro et al. [4] and Ore et al. [5] defines augmented reality as the technology that combines virtual information with a real scenario. Furthermore Schmalstieg and Höllerer [6] indicates that augmented reality meets three characteristics: a) combine real and virtual elements; b) interactivity in real time and c) alignment in 3D, helping to immerse the user in a computer-generated world, allowing an experience free of restrictions [6].

The advertising used in traditional channels has little impact, therefore, advertising campaigns with emotional content are developed, such as the case of school dropouts, which manages to sensitize users. Therefore Bassat [7] defines traditional advertising as the art of convincing consumers by visualizing the benefits of the product or service. However, at present it is difficult to convince the user through traditional means, which is why Alexander [8] indicates that digital advertising transfers traditional advertising to a digital world through social networks, websites, streaming and mobile applications, to which Cardona [9] recommends using the A / B test, which consists of developing two versions of a product to measure which one works better. On the other hand Pizarro et al. [4] presents advertising with augmented reality in the form of three-dimensional images, offering the consumer an experience that builds loyalty and generates a lot of expectation in the audience. Companies face a generation of customers who use information to purchase a product, which according to Godas [10] and Castilla et al. [11] are means that satisfy consumer needs and stimulate demand; and on the other hand, the services that according to Morales [12] and Chamilco et al. [13] is a process that consists of interaction activities between a good and the client. In this way, Bajaña et al. [14] indicates that companies should invest in advertising with augmented reality based on innovation. However, the target audience wants to interact with the product from a virtual environment to bring a personalized experience from home [15]. For this reason Pizarro et al. [4] proposes to use C ++ that allows the development of applications with three-dimensional animations and that works on mobile devices. Thus, augmented reality helps promote products and services in an innovative way, generating positive experiences through images on real environments, it also promotes electronic commerce by highlighting their characteristics [16].

In Spain, Moral et al. [17] carried out an investigation on new digital advertising strategies for young people such as advergaming that causes an emotional and affective immersion in the game, tabvertising that disseminates personalized and interactive advertising content specifically on tablets and smartphones and ARvertising where consumers interact with the product and personalize content from playful experiences with 3D resources, stating that advertising campaigns use augmented reality to increase sales in companies, allowing the user to experience the product through a virtual projection from mobile devices. Also in Ecuador, Bajaña et al. [14] investigated augmented reality in advertising, prospective for the Ecuadorian market that indicates taking advantage of the resources of the digital age without neglecting advertising content so that users experience a social connection with the brand, indicating that this technology is a trend with a future projection. In addition, it is revolutionizing and diversifying businesses, demonstrating the technological advance in first world countries. In Peru, Córdova and Jurado [18] in their research on augmented reality as an innovation in advertising media, they affirm that there was a greater acceptance of multimedia content, generating that 55.23% of the public recommend the use of augmented reality in the advertising because it helps interaction with images and videos.

In Spain, since 2007 little written advertising has been carried out due to the closure of 40% of existing newsstands due to the digital transformation and change in user habits; The custom of obtaining information through the internet has developed with the word "free" in mind [19]. In Mexico, fast food companies launched the "McBurritos" campaign, generating a negative impact on users for not respecting the country's traditional recipes. Advertising campaigns must be carried out under a user approach, that is, they must emotionally identify with the target audience. In Peru, clothing and accessories stores launched the #ModoCama campaign, causing discomfort in users because they describe an Afro-descendant person as messy and dirty. Whenever there is an advertisement, it must respect ethnicities, religions, races and sex [20].

Currently in the Peruvian market, companies seek to increase sales of products and services, they need to position themselves in the minds of consumers. They use advertising through traditional and digital channels, but do not have the desired impact to achieve an emotional connection with customers. The advertising campaigns present a moral and ethical content that manages to raise awareness about the issue at hand, but fails to persuade the customer to purchase the product or service. There is a new generation of consumers who use more information and want to interact with the product without physically having it to make a purchase decision. This type of consumer seeks to get closer to the product to verify its effectiveness and quality. In this sense, augmented reality and marketing are valuable allies for the growth of the brand. Therefore, the objective of this research is to describe the importance of augmented reality to promote products and services as an innovative alternative that captures attention and influences the customer's purchase decision, helping to generate a different and innovative experience to advertise products. and services through an emotional connection with people through mobile devices. This research describes the importance of augmented reality to achieve interaction with the consumer, personalize content, enjoy a new experience and stand out from the competition.

2. Method

The method used in the present investigation is of an experimental, quantitative and quasi-experimental design. The sample consisted of 60 Marketing managers from the retail sector, the technique used was the survey and as an instrument the questionnaire about the independent variable augmented reality with its dimensions of engagement, innovation and expectations of use with a total of 15 questions and the dependent variable promotion of products and services with the dimension level of innovation. The Likert scale was used as a data collection instrument, which was validated by the judgment of experts with a reliable questionnaire and Cronbach's alpha coefficient of 0.966.

3. Results

The questionnaire was applied to a total of 60 managers from the Marketing area, chosen for convenience. Based on the results, the following is presented:

Table 1, shows the results about the engagement dimension of the augmented reality independent variable, 20.00% indicate that the level of engagement is

Levels	Frequency	Percentage
Deficient	0	0%
Bad	5	8.33%
Regular	12	20.00%
Good	18	30.00%
Very good	25	41.67%
Total	**60**	**100%**

Source: Own Elaboration.

Table 1.
Level of engagement generated by augmented reality.

regular, 41.67% affirm that it is very good, 30.00% state that it is good and 8.33% considers it bad.

Table 2, shows the results about the innovation level dimension of the augmented reality independent variable, 28.33% indicate that the level of innovation is regular, 33.33% consider it good and 38.33% affirm that it is very good.

Table 3, shows the results about the expectation dimension of use of the same variable, 31.67% indicate that they personalize the brand or product, 28.33% consider that they make better use of technology, 23.33% affirm that it is innovative In the market, 10.00% consider that it offers a dynamic presentation of the product and 6.67% affirm that it generates interaction of the clients with the product.

Table 4, shows the results about the dimension level of innovation of the advertising campaigns of the dependent variable promotion of products and services,

Levels	Frequency	Percentage
Deficient	0	0%
Bad	0	0%
Regular	17	28.33%
Good	20	33.33%
Very good	23	38.33%
Total	**60**	**100%**

Source: Own Elaboration.

Table 2.
Level of innovation that augmented reality provides to the promotion of products and services.

Levels	Frequency	Percentage
They personalize the brand or product	19	31.67%
Better use of technology	17	28.33%
Innovative in the market	14	23.33%
Dynamic product presentation	6	10.00%
Customer interaction with the product	4	6.67%
Total	**60**	**100%**

Source: Own Elaboration.

Table 3.
Expectations for the use of augmented reality.

Levels	Frequency	Percentage
Deficient	9	15.00%
Bad	22	36.67%
Regular	6	10.00%
Good	11	18.33%
Very good	12	20.00%
Total	**60**	**100%**

Source: Own Elaboration.

Table 4.
The level of innovation of the campaigns to promote products and services of your company.

15% indicates that the level of innovation of the campaigns is deficient, 36.67% indicates that the level innovation is bad, 10.00% consider it fair, 18.33% think it is good and 20.00% say it is very good.

4. Proposal

Based on the results of the questionnaire, **Figure 1** shows the following model that proposes the importance of using augmented reality as an emerging technology to promote products and services, which help to evaluate the real state of traditional marketing campaigns and then apply the model and obtain results ideal, helping to generate a different and innovative experience to advertise products and services through an emotional connection with people through mobile devices.

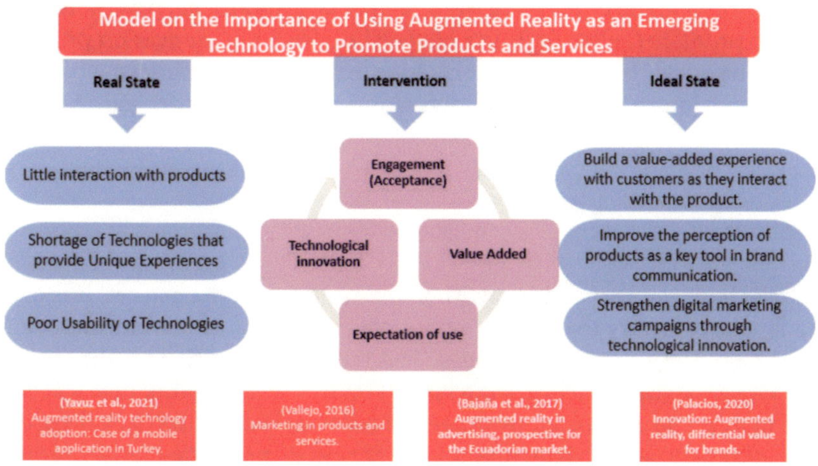

Figure 1.
Model on the importance of using augmented reality as an emerging technology to promote products and services. Source: Own elaboration.

5. Discussion

In **Table 1**, augmented reality generates a good level of engagement. Companies should not only use influencers or landing pages to capture the user's attention but also create an interactive experience that contacts and engages with them, such as augmented reality technology. Which agrees with Dwivedi et al. [21] who affirm that the design of the customer experience (CX) through emerging technologies such as augmented and virtual reality, voice-activated assistants and wearable technologies allow cultivating favorable customer engagement behaviors and improve consumer welfare. Bolton et al. [22] asserts that customers expect a fluid, integrative and holistic experience that is complemented with digital channels and people either by phone or in person, improving the overall customer experience.

In **Table 2**, the innovation generated by these types of experiences generates added value and a new reason for customers to try and adopt the company's new products. For Carlson et al. [23] digital and social media technologies that are integrated into people's daily lives provide new opportunities to interact and collaborate with brands in the innovation process. In this way, the innovation that augmented reality transmits is related to the audiovisual media that allows promoting products and services [24].

Table 3, indicates that augmented reality is an efficient means to market products or services. Which agrees with Hinsch et al. [25] that states that companies improve the perception of their products through augmented reality as a key ingredient in the communication of a certain brand. To improve the user experience, Han et al. [26] affirms that augmented reality applications are used in various industries to convey product characteristics and improve the way consumers think, feel and react about the decision of purchase. On the other hand, an analysis of the Company Ericsson [27] concluded that advertising with augmented reality could be a danger to improve the sales of products since the ads could be used as free versions of the products or services.

Table 4, shows that augmented reality is a technological innovation in digital marketing campaigns that develops more relevant and meaningful shopping experiences for consumers. Felix et al. [28] and Kamboj et al. [29] assert that digital channels are platforms that allow innovative marketing practices through social networks. On the other hand Muninger et al. [30] affirms that there is little guidance to promote the strategic use of social networks in innovation processes. While Leminen et al. [31] highlights machine learning and the internet of things as a perspective to improve the level of innovation of the Business-to-Business (B2B) business model.

6. Conclusions

With the huge number of users with mobile devices on the market, it is essential to use augmented reality to improve emotional connection or commitment, attract new customers and retain existing ones. This technology captures reality from different perspectives by superimposing virtual images on the reality that is seen from a certain mobile and supported by other technologies such as the Global Positioning System (GPS) help users to live unique experiences to interact with the brand through the information or entertainment.

Augmented reality is a technology that creates unparalleled, innovative and exciting experiences that contributes to closer ties between consumers and the brand. In addition to this, augmented reality filters in social networks and multiplayer gamification apps save time, travel costs and advertising thanks to the scalability and availability of the experience.

Companies use augmented reality as advertising and marketing tools to get the attention of consumers. In this way it is used in packaging, where the packaging of a product is used to display information. In events and activations, where positioning and good perception of the product are generated. In the catalog of products and services, where the products are brought closer to the user.

Acknowledgements

Special thanks to the National University of Cañete, the Research Group on Entrepreneurship and Innovation C@ñeteLab, its great professors, valuable students and collaborators for carrying out this scientific research throughout these years of study.

Conflict of interest

"The authors declare no conflict of interest."

Author details

Alex Pacheco[1*], Kevin Sánchez[1], Rosario Pariona[2], Nestor Cuba[1] and Bertha Larico[1]

1 National University of Cañete - UNDC, Lima, Perú

2 National Autonomous University of Huanta - UNAH, Ayacucho, Perú

*Address all correspondence to: apacheco@undc.edu.pe

References

[1] M. Yavuz, E. Çorbacıoğlu, A. N. Başoğlu, T. U. Daim, and A. Shaygan, "Augmented reality technology adoption: Case of a mobile application in Turkey," Technol. Soc., vol. 66, p. 101598, Aug. 2021, doi: 10.1016/J.TECHSOC.2021.101598.

[2] J. Bernal, L. Espinoza, E. Leo, L. Quiñones, and G. Moscoso, "Factores que influyen en el uso de las aplicaciones móviles en instituciones financieras mediante smartphone y su comparación entre las ciudades de Arica (Chile) y Tacna (Perú) - 3Ciencias," 3C Tecnol., vol. 7, Jun. 2018.

[3] L. Vallejo, Marketing en productos y servicios. 2016.

[4] P. Pizarro, D. Rodríguez, L. Rodríguez, H. Tong, and C. Villegas, "Plan de Negocios para una Empresa que brinda Soluciones Tecnológicas con Realidad Aumentada para Publicidad de Productos y Servicios," Universidad ESAN, 2016.

[5] J. Ore, A. Pacheco, E. Roque, A. Reyes, and L. Pacheco, "Augmented reality for the treatment of arachnophobia: exposure therapy," World J. Eng., 2020, doi: 10.1108/WJE-09-2020-0410.

[6] D. Schmalstieg and T. Höllerer, "Augmented Reality: Principles and Practice," 2016.

[7] L. Bassat, "El Libro Rojo de la Publicidad," 1993.

[8] L. Alexander, "The Who, What, Why, & How of Digital Marketing," May 2018.

[9] L. Cardona, "¿Qué es el test A/B?," Jun. 2018.

[10] L. Godas, "El producto. Tipos, atributos y diferenciación," pp. 116-120, 2006.

[11] R. Castilla, A. Pacheco, I. Robles, A. Reyes, and R. Inquilla, "Digital channel for interaction with citizens in public sector entities," World J. Eng., 2020, doi: 10.1108/WJE-08-2020-0377.

[12] V. Morales, "Evolución del concepto producto y servicio en la biblioteca: organización orientada al servicio e intensiva en conocimiento," e-Ciencias la Inf., vol. 8, no. 2, Jun. 2018, doi: 10.15517/eci.v8i2.30933.

[13] M. Chamilco, A. Pacheco, C. Peñaranda, E. Felix, and M. Ruiz, "Materials and methods on digital enrollment system for educational institutions," Mater. Today Proc., May 2021, doi: 10.1016/J.MATPR.2021.04.213.

[14] I. Bajaña, A. Zúñiga, C. Can, F. Meza, and A. Puris, "La realidad aumentada en la publicidad, prospectiva para el mercado ecuatoriano," vol. 10, 2017.

[15] K. Palacios, "Innovación: La realidad aumentada, valor diferencial para las marcas," Apr. 2020.

[16] J. Smith, "6 of The Best Advantages of Using Augmented Reality in Business Advertising," Business 2 Community, Oct. 2017. https://www.business2community.com/tech-gadgets/6-best-advantages-using-augmented-reality-business-advertising-01937379 (accessed Jul. 27, 2021).

[17] M. Moral, L. Villalustre, and M. Neira, "Estrategias publicitarias para jóvenes: advergaming, redes sociales y realidad aumentada," Rev. Mediterránea Comun., vol. 7, no. 1, 2016, doi: 10.14198/MEDCOM2016.7.1.3.

[18] M. Córdova and G. Jurado, "La realidad aumentada como innovación de soportes publicitarios en la captación de

estudiantes de educación superior," Apunt. Cienc. Soc., vol. 02, no. 02, pp. 101-109, Dec. 2012, doi: 10.18259/acs.2012012.

[19] N. Martínez, "Crisis en la publicidad. Causas, peligros y nuevos modelos," Oct. 2016.

[20] T. Forero, "Los peores fails de campañas publicitarias que nunca debieron ver la luz," Jun. 2021.

[21] Y. K. Dwivedi et al., "Setting the future of digital and social media marketing research: Perspectives and research propositions," Int. J. Inf. Manage., vol. 59, p. 102168, Aug. 2021, doi: 10.1016/J.IJINFOMGT.2020.102168.

[22] R. N. Bolton et al., "Customer experience challenges: bringing together digital, physical and social realms," J. Serv. Manag., vol. 29, no. 5, pp. 776-808, Nov. 2018, doi: 10.1108/JOSM-04-2018-0113.

[23] J. Carlson, M. Rahman, R. Voola, and N. De Vries, "Customer engagement behaviours in social media: capturing innovation opportunities," J. Serv. Mark., vol. 32, no. 1, pp. 83-94, Feb. 2018, doi: 10.1108/JSM-02-2017-0059.

[24] M. Fiadotau and M. Rajahonka, "Micro-trajectories: Small Firm Struggles at Boundaries between Audiovisual and Education Sectors," Emerg. Cross-innovation Syst., pp. 95-104, Apr. 2019, doi: 10.1108/978-1-78769-977-920191012.

[25] C. Hinsch, R. Felix, and P. A. Rauschnabel, "Nostalgia beats the wow-effect: Inspiration, awe and meaningful associations in augmented reality marketing," J. Retail. Consum. Serv., vol. 53, p. 101987, Mar. 2020, doi: 10.1016/J.JRETCONSER.2019.101987.

[26] D.-I. Han, M. C. tom Dieck, and T. Jung, "User experience model for augmented reality applications in urban heritage tourism," J. Herit. Tour., vol. 13, no. 1, pp. 46-61, Jan. 2017, doi: 10.1080/1743873X.2016.1251931.

[27] Ericsson, "The 10 hot consumer trends for 2016 - ConsumerLab - Ericsson," 2016. https://www.ericsson.com/en/reports-and-papers/consumerlab/reports/10-hot-consumer-trends-for-2016 (accessed Jul. 25, 2021).

[28] R. Felix, P. A. Rauschnabel, and C. Hinsch, "Elements of strategic social media marketing: A holistic framework," J. Bus. Res., vol. 70, pp. 118-126, Jan. 2017, doi: 10.1016/J.JBUSRES.2016.05.001.

[29] S. Kamboj, B. Sarmah, S. Gupta, and Y. Dwivedi, "Examining branding co-creation in brand communities on social media: Applying the paradigm of Stimulus-Organism-Response," Int. J. Inf. Manage., vol. 39, pp. 169-185, Apr. 2018, doi: 10.1016/J.IJINFOMGT.2017.12.001.

[30] M. Muninger, W. Hammedi, and D. Mahr, "The value of social media for innovation: A capability perspective," J. Bus. Res., vol. 95, pp. 116-127, Feb. 2019, doi: 10.1016/J.JBUSRES.2018.10.012.

[31] S. Leminen, M. Rajahonka, R. Wendelin, and M. Westerlund, "Industrial internet of things business models in the machine-to-machine context," Ind. Mark. Manag., vol. 84, pp. 298-311, Jan. 2020, doi: 10.1016/J.INDMARMAN.2019.08.008.